Mineral Rites

ENERGY HUMANITIES

Dominic Boyer and Imre Szeman, Series Editors

Mineral Rites

An Archaeology of the

Fossil Economy

Bob Johnson

Johns Hopkins University Press

BALTIMORE

© 2019 Johns Hopkins University Press
All rights reserved. Published 2019
Printed in the United States of America on acid-free paper
9 8 7 6 5 4 3 2 1

Johns Hopkins University Press
2715 North Charles Street
Baltimore, Maryland 21218-4363
www.press.jhu.edu

Library of Congress Cataloging-in-Publication Data

Names: Johnson, Bob, author.
Title: Mineral rites: an archaeology of the fossil economy / Bob Johnson.
Description: Baltimore: Johns Hopkins University Press, 2019. | Series: Energy
    humanities | Includes bibliographical references and index.
Identifiers: LCCN 2018032689| ISBN 9781421427560 (hardcover : alk. paper) |
    ISBN 9781421427577 (electronic) | ISBN 1421427567 (hardcover : alk. paper) |
    ISBN 1421427575 (electronic)
Subjects: LCSH: Energy consumption—Social aspects—United States—History. |
    Fossil fuels—Social aspects—United States—History.
Classification: LCC HD9502.U52 J654 2019 | DDC 333.791/30973—dc23
LC record available at https://lccn.loc.gov/2018032689

A catalog record for this book is available from the British Library.

*Special discounts are available for bulk purchases of this book. For more information, please
contact Special Sales at 410-516-6936 or specialsales@press.jhu.edu.*

Johns Hopkins University Press uses environmentally friendly book materials, including
recycled text paper that is composed of at least 30 percent post-consumer waste, whenever
possible.

To my little girl,
Ava Marie

Contents

Preface

A Postcard from the Birthplace of Oil

To begin, a postcard from Titusville.

This hand-colored postcard arrives to us from another time. It takes us back to the birthplace of oil, to Titusville, Pennsylvania, ca. 1859. The picture it portrays is of an undocumented oil fire that occurred forty or fifty years after the nation's first oil strike in the western Alleghenies, and its message is an odd one: "Will you accept this little card in place of a regular New Year's one. . . . I thot [sic] you might enjoy the scene from near my own house."[1]

That warm invitation to an industrial disaster, dispatched to a friend sitting in the midst of a cold Michigan winter, presents us with an interpretive dilemma. It contains none of the holiday cheer we expect to ring in the new year—to deliver us into a new season, a new future. No sleigh bells, no garland, no blessings—just this rousing oil fire, its carcinogenic smoke cloud rising, and a warm season's greeting on the back of the card.

A postcard from Titusville. *Reproduction from eBay*

What do we make of this invitation?

The aesthetics of this "oilscape" are not ours. Its mood—or what we might call, after Raymond Williams, its *structure of feeling*—comes from a time before climate change, before the current malaise, that is, before the full impact of industrialization had recolored the world in grey hues.[2] Its uncanny dialectic of turbulence and calm—of toxicity and leisure—is hard to process. For, on the one hand, we have this billowing cloud of carcinogenic smoke and waste rising boisterously from the center of the picture. It supplies, for all intents and purposes, the main event. On the other hand, the photographer has chosen a queer bourgeois formality and innocence to frame the picture. The sky is not industrial grey but rather hand-colored in pink pastels and robin's-egg blue, and the foreground showcases a curious innocence of young children squatting at rest in the grass. One young boy reclines in a white suit coat and boater hat as he leans affectionately toward the shoulder of a friend. Another perches on a tree stump in black britches and white knee socks, his shoulders hunched in resignation to the occasion. Knickers, bowties, and derbies, boys lazing on the grass—this is an odd framework for an industrial disaster.

A burning oil tank with children posing in foreground. *Reproduction from eBay*

The message given to us resists the neat political and symbolic boxes we have for making sense of the waste and worry that an oil fire signifies. It refuses to be stuffed into a neat paradigm of melancholy or shame, challenging us instead to embrace a Victorian confidence, a different way of being in oil. Far removed from the scene are the smog of Los Angeles, yesterday's shrimpers collecting compensation payments along the Gulf Coast, and the Larsen Ice Shelf collapsing into the Weddell Sea. These episodes in climate change are part of a future that has not yet been written into the landscape or entered into the historical record. Such worry is put off for another day. These leaky clouds still look like progress.

Context is everything.

The Fossil Economy

The 1859 oil strike that occurred in Titusville, Pennsylvania, marked the nation's first oil boom. It was the original event that set us on the path to becoming a modern petroculture, even if the West had already gone down the rabbit hole of modernization and global warming in its earlier embrace of coal. Petroleum in the West, however, begins here. Titusville was preceded only by the contemporaneous boom in Azerbaijani Baku a decade earlier on the other side of the world. Oil had a history before Titusville, but it was a marginal one. Iroquois Americans harvested petroleum from the region's Oil Creek for use as a tonic and grease. Non-native locals also harvested oil to prepare medicinal emetics, and elsewhere across the preindustrial world, crude oil located near the surface was applied to pipes and ships for waterproofing and, in a few instances, used as a quickening agent to create firestorms in war.[3] But until Titusville, oil was only a marginal part of the ecosystem, a quiet resource, a historical bystander rather than participant. It had yet to join coal in the world's coming jubilee.

Tapping into the Petrolia reserve changed all of that. From 1859 forward, these subterranean forces became terrestrial, and the ecology of modern life become increasingly primeval. Crude oil, this residue of fossilized history, of "compressed time," as Jennifer Wenzel puts it, was pumped out of the ground with great speed and rapidly integrated into a transatlantic, and ultimately global, marketplace of commodities that already included colonial consumer goods like sugar, tea, and cotton.[4] For the next fifty years, petroleum from fields across the globe, such as Petrolia, Baku, Pico Canyon in California, and North Sumatra's Telaga Tunggal, lit up the transatlantic and transpacific

worlds, bullying aside the world's whale oil industry and providing the kerosene for light used in kitchens and bedrooms, factory and office interiors,
and bustling urban street corners. Petroleum accelerated and altered the
pathways into which coal had already steered us within the fossil economy.

But at the turn of the century, when this picture was taken, possibly
when lightning struck an oil tanker in 1894, petroleum was still of a different order. It was a commodity, yes. It was an energy source, sure. And it was
the developing world's preferred illuminant, used for lighting up homes and
offices. But it was not yet a source of *power* or *work*. The combustion engine
was still in its prototype, and no one could have anticipated a raucous mineral
economy characterized by 2-ton personal vehicles hurling tens of millions of
people across ribbon-like highways or 220,000-ton cargo ships transferring
resources and goods in a global system of oceanic highways. Nor could we
have anticipated the accumulated costs of these new mineral dependencies,
what Bill McKibben has designated to be "the end of nature" that came from
unchaining carbon from its underground moorings.[5] At the turn of the century, when this picture was taken, it was easy to think of oil as a form of light,
of lux, of enlightenment, and the scope of its environmental consequences
(although not its economic impact) as localized and containable.

This postcard from Titusville reminds us, in other words, that not so
long ago we could still gather the nation's middle class around an oil fire with
a certain degree of innocence—that it was not so far back in our history that
the romantic world of Jane Austen, with its foreground of green meadows,
picnic blankets, and thick trees and its background of ugly colonial subsidies
of slave labor and wage slavery, had ceded to this hyped up and spastic one
of combusted carbon, with its new liberties, new costs, and new enslavements. We are only a few generations gone.

The words of historian Jean-François Mouhot are of some value here:
"We have arrived at the present situation (mostly) in good faith, with the
conviction that modernity would bring the masses freedom from toil, and
without any chance of knowing the climatic consequences of our burning of
fossil fuels."[6] Carbon's most aggressive boosters, from the predatory John D.
Rockefeller to the equally destructive Koch brothers, have known better and
operated with a systematic cognizance against nature and humanity, but the
rest of us might find some reprieve in knowing that the average consumer has
until recently participated in carbon's jubilee with an unmetered innocence.

The Deep Ecology of the Modern Soul

There is, in other words, a lesson to be learned here.

This postcard invites us into forgiveness—to absolve ourselves from having gotten ourselves into this predicament. It gives us permission to move forward with renewed conviction out of carbon's moment and away from the unanticipated consequences of the choices that brought us here.

Even so, this curious postcard points us to a second lesson. It challenges us to confront what has for too long been a distractedness in attending to the externalities of our lives. Suit coats and black ties, cameras and awkward smiles, GDPs and declarations of progress, our world's insistence on optimism and an enduring bourgeois insistence on form have for too long worked to silence these clouds that sit so portentously in the background to our lives. Most of us will never travel, dressed in our Sunday best, to the coal mines, oil platforms, and boiler rooms of the world to see what lies beneath the Western standard of living, but we know, deep down, that the logic of the fossil economy ties pleasure to pain, comfort to suffering, and present health to future injury.

This memory from Titusville gives us, that is, a glimpse into the deep ecology of the modern soul and a premonition of the impending collapse of fossil capitalism.

Acknowledgments

I wrote this book for my daughter Ava—for her future and for memories to be made. She is a beam of light, a little steady radiance, in the drift of the world.

This book, like all creative projects, was written on the backs of many other people. I am fortunate to have had the support of my family along the way. My mom, Sherry Johnson, and my dad, Robert Johnson, did everything in their power to give me a liberal arts education and to crack open the world's possibilities for me. My parents still mean the world to me, and I can only hope to pay them back with the same dedication, hard work, and love for my own. This book is a small repayment for their love and sacrifice. My brothers, Kevin Johnson and Mike Johnson, are never far from me. They are my friends, and they served as role models throughout this process, reminding me of the hard work it takes to make something out of the world, to get a little something done, even as they always make room for our family in the maelstrom of life. My sister, Nikki Andrews, is not far from this book either. She holds things together in an all-too-spinning world by trying to keep the family corralled, to give us some center across the miles. So, thank you too, Nikki. Likewise, I feel grateful to Dan and Jean Gallo and their family for their conversations, kindnesses, and interest over so many years and over so many nights with a glass of wine in hand. I miss them. And, finally, I want to say thank you to John and Karen LaFayette who, years later, manage across the miles to toss in their love and support as a key ingredient in this process.

A number of my professional debts are immediate. My friends Eileen Luhr and Colin Fisher deserve special recognition for serving as sounding boards from the start to the end of this project—and, more specifically, for commenting on draft chapters along the way. Having them there to rein in my excesses and keep me attentive to the bigger historical context was invaluable. Similarly, my friend and colleague Julie Wilhelm gave me key advice on a later chapter of this manuscript. I couldn't have written "How to Read a Novel in Light of Climate Change" without her suggestions and without the advice of a few other friends in our literature department, Amina Cain and John Miller among them. Thank you. My friend and colleague Duncan Campbell likewise offered important feedback on the chapter "Energy Slaves," which needed a critical eye before it went to press. I also want to acknowledge my colleagues in the Petrocultures Research Group: they are the conversation that I dropped into, inspiring me, challenging me, and helping me to see that we are all part of a collective enterprise that is evolving together. Within that group, Imre Szeman, coeditor of Johns Hopkins University Press's new series, Energy Humanities, deserves special notice. He has played a key role in building the sandbox that brings many of us in the energy humanities together, including such inspirational thinkers as Matthew Huber, Graeme MacDonald, Sheena Wilson, Jeff Diamanti, Brent Bellamy, Darin Barney, Mark Simpson, Jordan Kinder, and a constantly growing group of friends and colleagues. Finally, I'd like to say a special thank you to Jason Twilla, Josh, Ty, and Peter at Influx, both for their conversations and for loaning me real estate in the often lonely pursuit of a sentence.

A version of chapter 1 appeared as "Embodiment" in *Fueling Culture: 101 Words for Energy and Environment*, edited by Imre Szeman, Jennifer Wenzel, and Patricia Yaegar, pp. 124–127. Copyright © 2017 Fordham University Press. A version of chapter 3 appeared as "Energy Slaves: Carbon Technologies, Climate Change, and the Stratified History of the Fossil Economy" in *American Quarterly* Volume 68, Issue 4, December 2016, pp. 955–979. Copyright © 2016 The American Studies Association. Thank you to both publications for permission to include that material here.

My editor at Johns Hopkins University Press, Matt McAdam, also deserves recognition; as do my two external reviewers who read the manuscript with a spirited eye and professional care. The book is better because of you.

Mineral Rites

Introduction

The Mineral Moment

This book is an archaeology of the present. It unearths a buried set of stories about the origins of modern life that center on the pivotal, if sublimated, role that fossil fuels have played in the growth of capitalism and in fueling our affective attachments to modernity. It recovers, in this respect, a genealogy of the modern self, one oriented to a postcarbon future, by demonstrating how our bodies, minds, identity, nature, reason, and faith are energized by, given life by, an infrastructure of carbon flows where we are fleshed out in everyday mineral rites, fossilized rituals, that imbue our sinews with muscle memory, provide material for our imagination and senses, and shape our expectations about being fully human in the twenty-first century. In short, this book proposes an alternative history of modernity that returns our attention to the dialectic of materiality and consciousness, body and affect, that has taken shape under the terms of fossil combustion during the past two hundred years—or, to push the matter further, under the terms of "fossil capitalism" during that time.[1]

Genealogy, we know, is grey, meticulous, and heterogeneous—defiant of origins, possessive of its singularity.[2] Thus each chapter of this book comes at this project from its own uncanny perspective, narrowing in on an artifact, trope, or ritual in the social life of fossil fuels: a chunk of coal recovered from the *Titanic*, a day spent in a hot yoga studio, a postcard sent from the nation's first oil boom. From that particularity, each chapter maps out, piece by piece, chunk by chunk, the ontological depths of the fossil economy so as to raise to the surface, to make visible, the material, sensory, and emotional substrata through which carbon enters into our lives—into our labor and play, our bodies and consciousness, albeit in ways stratified by class, race, religion, gender, and nation.

The title of this book, *Mineral Rites*, refers to the various rituals—from the morning shower to online shopping, from freeway commuting to breathing near a petroleum refinery—by which we naturalize these energies, taking them into our bodies in ways we recognize and ways we don't. The subtitle of this book, *An Archaeology of the Fossil Economy*, nods to the type of genealogical work that runs methodologically through it.

The main argument is that modern lives—male and female, white and black, rich and poor, American and Bangladeshi, Christian and Muslim—are conditioned by a global infrastructure of carbon flows that saturate our habits, thoughts, and practices but that tend to be socially barricaded behind No Trespassing signs and cordoned off, symbolically and psychically, from fantasies of the American Dream and its global permutations. Integral to that argument is the assumption that this embodiment of fossil fuels, including our affective attachments to them, is a highly stratified affair, one that cuts along differential axes of power.

At the center of this argument thus stands a paradox first developed in my previous book, *Carbon Nation*. That paradox runs something like this. Fossil fuels are the source of health and opportunity, fertility and reproduction, in the modern world, and they prop up the rich emancipatory qualities that many of us, especially in the upper and middle classes, expect from modern life in the West (including easy mobility, materialism, economic growth, and a diversified palate). But they are also and simultaneously the fuel for widespread social injury and limited horizons, for impotence and decline—a prime mover of fear, neurosis, and terror for those living downwind of power and for those of us staring into an unknowable future.

To be more precise about this, fossil fuels are, on the one hand, the leverage of modern life, the fulcrum on which modern populations pivot. To understand fossil fuels in this sense means understanding that they are not simply "fuel" for our personal vehicles, furnaces, and air conditioners—a source of heat and propulsion—but also a material supplement, an excess of the organic, that grants to modern economies a larger and deeper degree of resource flexibility with which to meet humanity's basic Malthusian needs. We think we know fossil fuels, but we tend to size them up insufficiently. The term *fuel* appears as an impoverished signifier; *energy* a term that merely nails jelly to the wall. Fossil fuels are more than what our language permits us to see, and thus we need a proper *energy heuristic* to tease out the hidden material functions they perform behind the scenes to birth us into a second nature.

There are at least five functions, or modalities, through which we embody fossil fuels and incorporate them into our daily routine and environment. These include:

- *Ambient energy*: carbon generates the habitat and habitus of home through lighting, heating, and air conditioning (i.e., HVAC systems), remaking the setting and mood of modern life as well as the impression of security in it.
- *Congealed energy*: carbon creates today's hardened and vertical material environment by expanding the heat supply for blast furnaces, gifting us a tough, soaring, and sprawling world of steel, glass, and concrete, playing a critical role in housing and infrastructure.
- *Polymerized energy*: carbon regrains modernity's textures by serving as the fuel stock for a synthesized and pliant world of fiberglass, nylon, polyester, PVC (polyvinyl chloride), and Lycra, plasticizing the world and expanding its fiber supply for the purposes of clothing, furnishings, architecture, and consumer goods.
- *Embodied energy*: carbon reinvents the rules of species reproduction, elevating carrying capacity (i.e., population) and remaking the bioenergetics of food security by both taking us outside of nature's nitrogen and phosphate cycles (e.g., artificial fertilizer and phosphate mining) and providing the background refrigeration and propulsion needed to fatten granaries in a complex global food system.
- *Propulsive energy*: carbon reinvents labor (i.e., work, force, mobility) by supplying mechanical energy in motors, engines, and turbines that propel us beyond the somatic economy of the human body, permitting us to escape an austere preindustrial economy of slaves, servants, serfs, yeoman, hunters, and artisans, driving and accelerating industry, transport, and even leisure.

In these diversified roles, fossil fuels perform an *eco-logical* magic. They flex our resource base, gifting us with a more generous set of options for taking what we need from nature. In effect, they emancipate us from the past, from the older biological constraints of forest growth, soil fertility, draft horses, and slaves that had previously imposed hard limits on reproducing the essentials of life (fuel, clothing, food, and shelter). By permitting us to move beyond the forest—to circumvent the earth's nitrogen cycle and soil constraints and to escalate our labor productivity beyond the capacity of physical bodies—fossilized carbon rapidly became the foundation for more than a century of optimism about the future. Today, fossil fuels support a global carrying capacity that rises skyward toward ten billion people, looks back on two hundred years of rambunctious growth in global GDP (gross domestic product), and squints hopefully forward, albeit with some trepidation, to a future that does not end in preindustrial austerity or system collapse.

The world's demographic horizons, its potential for life, and the possibilities of its politics all derive from this zombie-like reactivation of fossilized life brought up from underground.

On the other hand, these mineral fuels operate simultaneously as the agents of unrepaired harm, inequality, and evasion. The fossil unconscious—this repressed underside of the fossil economy—is populated by terrors, neuroses, and pain. To speak excitedly, and blithely, as many of us do in the West and the developed core, in tropes of emancipation, opportunity, and progress, is misguided. Such rhetoric reflects a strong bourgeois bias and an unrepentant presentism that naturalizes the vantage point of comfort and endorses a cultural form of fossil imperialism. Such a perspective, long promoted by industry boosters, state actors, and the many beneficiaries of the fossil economy, supposes that these energies have been the main driver of an Enlightenment project originating in the West that produces social mobility, physical emancipation, and material comforts delivered through science, engineering, and vision.

Yet, as we now know, this highly selective representation of the world we live in produces major blindspots and distortions. Opposite, under, and inside of this privileged way of being in fossil fuels are the other partially concealed, and marginalized, experiences of injured peoples, spiraling penury, and broken ecosystems. The fossil unconscious is a register of this misshapen world that buckles under the strains of radical resource extraction and unequal colonial and quasi-colonial practices.

That underside, moreover, is not simply a repository of past and present pain. It is also future-oriented. Combusting our way into modernity has produced not only a Janus-faced landscape of mobility and immobility but also a new environmental stage for us to perform on, a destabilized climate that veers away from the relatively predictable patterns under which *Homo sapiens* as a species flourished during the Holocene. We live in a new planetary context, even if we are still learning about that context. Extreme heat waves, polar vortices, and the normalizing of drought: these are the barely repressed symptoms of a high-energy world that is producing system failure on a planetary scale—and transferring worry to the future. To steal a phrase from the literary theorist Rob Nixon, we are watching our future be degraded through a type of "slow violence" that does not register in the old ways.[3]

This is risky because reproducing modernity, even in its plans B and C, presumes climate and water conditions more or less akin to those that allowed for the expansion of agriculture and the creation of a global hydrological infrastructure over the past ten thousand years. The weirding of the earth, or as Bill McKibben suggests, this new "Eaarth," is no longer promising that sort of predictability.[4] Fear today lies in the nonlinear. The fragile hybrid crops of the Green Revolution, the resurgence of organic agriculture, the promise of GMOs (genetically modified organisms), and even the lingering practices of hunting and gathering may not weather the unfamiliarity of extreme climate volatility.

And so to our energy heuristic we must add a sixth term to capture this id-like other, a term to name the destructive impulse, the death drive, internal to modernity. We might call it simply *entropic energy* in recognition of the fact that carbon's combustion works simultaneously to accelerate disorder: to degrade nature's complexity through emissions and waste (i.e., CO, CO_2, and garbage piles of disposables) and to arm us with the fossilized weaponry and other industrial-grade technologies that shake up social systems, disrupt life patterns, and degrade the ecosystems that *Homo sapiens* depend on to thrive as a species.

Entropic energy (this palpable register of a repressed fossil unconscious) points us back, that is, to the damages and leakage of modernity, to its externalities and its waste products. It thus returns our attention to the heavy-duty scrubbing we have done to keep that leakage off stage and out of our minds, realerting us to modernity's internal entropy, to the presence of injury, so that we might accept the ambivalence of the world we have created and thus better anticipate our prospects going forward.[5]

We have today some hard thinking to do. It might be true that we humans have a wide latitude in our capacity to construct our demographic horizons and to adapt to material conditions, but even so, demographics—this art of feeding, fueling, sheltering and clothing people—will provide a bottom line that matters. It will have the final say.

But for now, we trade the future for the present, social excess for social justice, whisking away objections with balloons and confetti. Despite some very good science and a capacity for good sense, we just can't seem but to urge ourselves onward to invest in the project of modernity and to believe that it is simply not finished, that it just requires more fuel. Yet we know that this is the same thing as making peace with our children's downward mobility, ensuring they will live and give birth amid radical worry and stress.

What has made this paradox palatable up until now is that the fossil economy has done real work. It has underwritten two centuries of middle-class expansion and the increased material security of many people at home and abroad. In North America, for instance, combusting carbon, under the terms of fossil capitalism, has been a genuinely good deal for those of us who climbed out of an austere yeoman economy and into a degree of relative privilege and security. Homes that are 2,600 square feet in size (American average), with central heating and climate control, a 26.4-minute commute (American average) without one having to break a sweat, a secure, diversified, and excessive food supply (2,000–3,000 calories, American adult male average) transported through a global infrastructure of calorie production. Such pleasure and privilege are in historical terms not something to laugh off.[6]

Moreover, the mineral economy has operated to reinforce self-interest by performing a certain magic on its beneficiaries. That has been true whether we look at privilege and comfort under fossil communism in the Soviet Union, under fossil fascism in Germany, or under fossil capitalism in the West and now globe at large. Each of these permutations of the fossil economy has managed to cordon off the experience of pain from its main pleasure centers by exiling its externalities, if not always successfully, to far-off places—to Russia's remote Magnitogorsk at the edge of the Ural Mountains, to the water-tight stokehold of the modern steam liner deep below the promenade, or to a depressed Appalachian hollow where disability and economic morbidity are written off as someone else's troubles. We have made sure, that is, that the pain points of the fossil economy have always been cut off socially and psychically from the joy centers of its marquee geographies, whether the petro mansions of Russia's oil-igarchs or the segregated pastoral communities of the postwar American suburbs.

But what does it mean to sever the moral ligaments of the world we live in? And what does it mean to leverage life today on tomorrow's suffering?

This book seeks an answer to those questions by examining our mineral rites, by picking them up, looking under them, and circling around them so that we can get a more accurate feel for the eroticism they produce and the socioecological costs they reproduce. It focuses particularly on the experience of fossil capitalism in the West, as its rise is today's main storyline. It need not be read in sequence, as each essay is an autonomous meditation on our fossilized lives, although the chapters move forward with some accumulated momentum.

Each of these chapters comes at this project in a unique way by denaturalizing the subject of modernity, taking apart our ways of knowing and being under fossil capital, and subjecting them to scrutiny. Stylistically, they eschew chronology and linear narrative and forego an exclusive reliance on sometimes sterile academic forms to propose a phenomenology of writing that better registers the barrage of life thrown at us each day and that invites us into a bricolage aptitude more suitable to survival under modernity.

The first chapter, "Mineral Rites: The Embodiment of Fossil Fuels," examines how the middle-class self is choreographed by an infrastructure of carbon. To dramatize that point, it turns to the strange and hyperbolic ritual of contemporary hot yoga, using that high-energy site of fossil consumption as a hyperbole for the modern condition at large. It demonstrates how the bourgeois self takes its form in such privileged sites of combustion through an erotic interplay of body, consciousness, and fossilized energy inputs. In so doing, it gives us a portrait of the class dialectic that ties modern self-realization and satisfaction to a sublimated infrastructure of energy flows that generate, as its waste, the derealization, the disability, the labored breathing, and the ecological disarray of modernity's other.

Chapter 2, "Carbon's Social History: A Chunk of Coal from the 1912 RMS *Titanic*," explores the social life of fossil fuels. It returns us to fossil capitalism's purest, most Platonic achievement—the *Titanic*. Centered on a chunk of coal recovered from that luxury steam liner, the chapter takes us back through the commodity chain that drove this monument of fossil capitalism to see how an infrastructure of high-grade coal fueled the *Titanic*'s starkly stratified world of pleasure and risk—of Turkish steam baths and sweaty stokeholds, of hungry Welsh miners and engorged elite passengers, all dialectically bound to one another in this microcosm of fossil capital.

Chapter 3, "Energy Slaves: The Technological Imaginary of the Fossil Economy," unpacks the fossil imaginary, or to be more precise, it unmasks the technofundamentalist dispositions of modernity that lead us to believe that fossil fuels (and the machines they empower) have emancipated us from nature, from labor, and even from history. Focused on the curious (and quickly forgotten) corporate unveiling of a black robotic slave in 1930 by the Westinghouse Corporation, it traces out the trope of technological servitude from its early appearance in the West's language of mechanical slaves during the Industrial Revolution through today's political debates over the role of "energy slaves" in climate change.

Chapter 4, "Fossilized Mobility: A Phenomenology of the Modern Road (with Lewis and Clark)," denaturalizes modernity's background culture of time and space by recovering the phenomenology of the modern road and modern mobility. To achieve that, it takes us on a tour across the continent with the archetypical trekkers Lewis and Clark, who recorded in great detail what it meant and felt like, in social and sensory terms, to move one's body across time and space without prehistoric carbon, without the combustion engine, the diesel engine, the railway, or the commercial jet. This chapter thus interrogates our second nature by making visible the strangeness, including the unknown risks and dangers, of today's uninterrogated culture of automobility.

Chapter 5, "*Coal TV*: The Hyperreal Mineral Frontier," steps into today's hyperreal media environment to explore the difficulty of seeing fossil fuels, and our dependencies on them, accurately in the context of systematic media distortions. Focused on a reality television show, *Coal TV*, that followed the lives of hard-rock coal miners in a small and otherwise sleepy town in West Virginia, this chapter shows how today's media culture has worked to naturalize fossil capitalism and to minimize the costs of our energy dependencies by accommodating them to comfortable liberal, and now neoliberal, tropes of agency and emancipation.

Chapter 6, "Carbon Culture: How to Read a Novel in Light of Climate Change," returns us to the energy unconscious that sits just beneath the surface of our culture. It applies to our literature the energy heuristic I have outlined and reflects on how fossil fuels drive our cultural texts below the register of plot, showing up as the ambient energy, propulsive energy, congealed energy, polymerized energy, embodied energy, and entropic energy that supply the elemental ground to our fiction. In doing so, it explains how we can begin to reskill ourselves for reading and writing in light of climate change by becoming attentive to the modalities by which a fossil infrastructure asserts itself both materially and ideationally in our cultural life.

In the Epilogue, this book offers a reflection on the question of temporality in light of fossil-induced climate change by deconstructing three of today's big metanarratives, the Anthropocene, Big History, and the Capitalocene, asking if there might be more productive strategies for writing the history of the now so that it better fits within today's prevailing structures of feeling that can mobilize us to action.

Terms and Definitions

Several key terms knit these essays together, and it is thus worth taking stock of them.

The first term—*fossil economy*—is developed in my previous book *Carbon Nation*, and it is a term that appears periodically in the work of other scholars, perhaps most notably that of human ecologist Andreas Malm.[7] I utilize this term as shorthand for saying that we have lived, since the mid-nineteenth century, under a different set of ecological and economic assumptions than we did for the previous 200,000 or even 2.5 million years of human history, and that we can thus mark a relatively sharp break between what we call the modern and the premodern worlds (and what we categorize as the industrial and preindustrial experiences). For ecological anthropologists and energy historians, this claim will appear to be old wine in a new bottle. Leslie White, William Catton, E. A. Wrigley, Rolfe Sieferle, Kenneth Pomeranz, John R. McNeill, and Edmund Burke have all argued previously, albeit in different ways and to different ends, that fossil fuels mark the break between the then and the now.[8] They each explain how new energy flows from below the soil altered the ecological calculus by which we could advance the world's carrying capacity and by which we could engineer radical growth in labor productivity and global GDP. That is, it refers us back to the moment when economics was the world's dismal science—and to when it was realigned to today's boosterism.

For my purposes, this term, *fossil economy* (which refers to a mode of production that emerged two hundred years ago and gradually expanded from a handful of textile mills in the West to the globe at large, affecting over that time pretty much everyone in one way or another), allows us to frame the world we live in as being of a single piece, a synchronic slice in the human experience. It is a way of defining the modern moment, its origins, its substance, and its finality.

This terminology warrants a curious caveat. Fossil fuels are, of course, not technically minerals. They are organic matter composed of plants and animals, snails and ferns from the long past that became fossilized (or mineral-like) under geological pressure in the absence of oxygen. To be a purist about it, they are thus properly classed with other organic life forms, past and present, that contain carbon, and thus distinct from iron, copper, gold, silver, aluminum, and uranium—these truly inorganic elements that science designates as the real minerals. But the world is socially constructed, and reality doesn't get made in the purity of the glove box. Things, as we know, get messy outside of the laboratory. Coal, oil, and natural gas have long been classified not as organic matter but as minerals in legal, economic, and political discourse, if not in the natural sciences. That is to say, over the course of time, we have decided that oil and coal have less in common with a hungry badger or a blooming petunia than they do with aluminum and iron. Hence, drilling and mining on federal lands were first clarified under an act we call the Mineral Leasing Act of 1920, which lumped fossil fuels into the category of the mineral and promoted their development alongside other inorganic resources like phosphate and sodium. Consequently, we use the term *mineral rights* to designate property regimes in the fossil economy, and hence, the title to this book, *Mineral Rites*.

A second term, *fossil capitalism*, serves a specific purpose in this book. That term derives from Malm's *Fossil Capital: The Rise of Steam Power and the Roots of Global Warming*, which contends that the fossil economy is by origin and subsequent developments a distinct form of capitalism. The fossil economy started, Malm tells us, amid the drive and ethos of industrializing capital in the entrepreneurial textile mills of England, and it served in that context not as the solution to material obstacles nor as an answer to bottlenecks in technological inefficiencies but rather, and quite self-consciously, as the technology by which aspiring capital could increase its control over labor. Malm reminds us, in other words, that the complex of coal and steam was coaxed into life by certain historical actors and for certain class-based reasons.[9]

That argument is compelling for two reasons. First, Malm explains that mechanization in the textile mills occurred not in the absence of cheaper forms of labor, as we might expect, but rather amid a surplus of labor. Mill owners began to use mechanical energy, whether supplied by water or coal, he says, to circumvent their reliance on the working poor, whose output was not rationalized and who systematically pilfered from the profits of the mills to supplement their meager income. Second, he tells us that coal and steam, as opposed to waterpower, introduced specific managerial advantages for industrial capital. In particular, it gave investors the ability to relocate factories away from a handful of interior waterways, where waterpower was sited, and to capitalize on a different class of workers in cities and towns who were not yet organized. The effect was quite deliberate: steam manufacturing allowed capital the mobility to undercut the strength of men's spinners' unions and to do an end-run around workers' control. It gave capital a more pliant labor force, and it enabled a rationalization of its inputs and output from the top. It allowed modernity's first class to take shape.[10]

This arresting meme—fossil capitalism—captures the dominant strain in the West's experience with fossil fuels and in the globe's current experience with them, and it gives us something to swing at. But it too requires a brief caveat because the argument both exercises a sort of originalism, or birtherism, that colonizes the fossil economy's subsequent history and a certain presentism that draws too neat a line from today's impasse to fossil capital's origins. By casting off fossil communism in Eastern Europe and fossil fascism in Germany, Italy, and elsewhere as side notes deserving of only a short paragraph, this argument casually throws off much of the history of the twentieth century. It renders big events like World War II a historical oddity and the Berlin Wall a minor event, while it writes the collapse of communism into its DNA and treats the history of fascism as a thing of the past. These too were fueled by, and they were about, the political command of carbon. The wild success and current victory of fossil capitalism over its competitors does not make fossil capitalism the only, or even always main, storyline, even if it appears that way in hindsight. But even with that caveat, Malm's intervention meets the cut of the pragmatists' razor.

A third term used throughout these essays is *fossil unconscious*. I put forward that term to refer us to the raw underside of the fossil economy, to the alter ego of fossil capitalism, as a way to take us back into our world's infrastructure, into production, and into that barely concealed world of toxic seepage and collective human casualties that has always been part and parcel of the mineral order. The fossil unconscious is thus two things at once. It is first and foremost a material thing, rooted in radical ecological degradation and the stratified experiences of those who live at the end of carbon's commodity chains. But it is also a psychic matter, a mental maneuver that represses the presence of infrastructure, production, and their human and environmental costs. The fossil unconscious thus refers us back, in this second sense, to those old phrases that were once meant to capture the social dialectic of industrialization's pleasure and pain, phrases like "how the other half lives," "the other America," or "the other side of the tracks"—although, to be sure, it adds to these old concerns a new ecological dimension that refers us also to climate change. The fossil unconscious traces back, that is, to our economy's sacrifice zones, its spaces of economic morbidity, its demented ecologies, and its generation of blocked desires. Its repressive element, which is both psychic and social, encourages us to look away from the alleys, favelas, and underpasses of the fossil economy and away from the crisis that is in front of us. The fossil unconscious is modernity's destructive and antisocial impulse; it is the substance of the modern world's id and a register of that entropic energy that thrusts us forward.

These terms are all reaching toward something. They are reaching toward a temporality of the present, toward a way of slicing history into a meaningful synchronic piece, so that we might get a better look at it and be able to figure out when it was that this "we" began and why this particular version of "we" has gotten us into so much trouble.

Interventions and Threads

By way of closing, let me state up front and more directly a few of the main interventions of these essays.

First, and most obviously, is their focus on how fossil capitalism orchestrates human nature (e.g., bodies, minds, and souls) and nonhuman nature (e.g., soils, forests, and waterways). These essays show how that orchestration occurs through a deep and mostly uninterrogated culture of *mineral rites*, which is, in turn, organized by a transactional marketplace fueled by carbon. To that end, the essays in this book each take inventory of fossil capital's presence, immanence, multiplicity, its little scattered referents, evident everywhere but named only in the breach, to bring it out of the shadow and into the light. They reveal fossil capital to be more than background, more than second nature—to be a historical event and thing of agency, mobilized, reaffirmed, concretized, and validated day in and day out through a praxis of petroleum and through a way of being in bitumen. By returning us to the world's coal fields, its polyvinyl chloride factories, its hot yoga studios, its dining rooms, its automobile interiors, and its oil refineries, from Los Angeles to Beijing, the following essays give us a feel for how today's fossil economy, organized by capital and engorged by cheap energy, aligns us to capital's blueprint for life, people, and planet.

A second and corollary concern of these essays is the ethics of life under fossil capital. Today, we know that empathy, a core disposition of any ethical system, is in short supply, relegated to the sidelines of history's main event, assumed to be marginal to modernity's real movement. That leaves us with some need to reinstate the centrality of human relationships—of community and solidarity—in our historiography so as to point us toward healing and suturing. To that end, these essays offer a disruptive optics focused on capital's knotted skein of circulations, including all of those seemingly agnostic market operations that twist and distort living today into the systematic abuse of subaltern bodies and the planet's ecologies. The simplest point to be made in this respect is that fossil capital today systematically pits producers against consumers while erasing (i.e., burying inside a global system of impersonal circulations of wealth, labor, and commodities) their solidarity with, and obligations to, one another. These essays provide some counterbalance to that estrangement by walking us back through the world's commodity chains and through their correlating chains of signifiers. By doing so, they highlight the bittersweet and confusing entanglements of living in a world where freedom, health, and mobility generate disability, immobility, and morbidity somewhere else, for someone else, in ways stratified by class, race, gender, and region. If today fossil capital cries out that these relationships are already accounted for by the mechanism of the dollar exchange, these essays suggest that they aren't—and that we all, deep down, really know better.

A third concern of these essays is the importance of politics under fossil capital—in particular, the importance of understanding that a class of architects stand directly and indirectly, wittingly and unwittingly, behind this project. The following essays call attention to the fact that a small circle of financial actors, which we proverbially call the 1%, combined with a panoply of entangled state actors, play an outsized role in directing modernity's project.[11] Throughout these essays, we catch these cyclopean giants restructuring the rules of the game that we are all playing as they reorganize transatlantic shipping lines for the *Titanic*'s maiden voyage, open and close dirty strip mines and oil refineries on the periphery, trade stocks behind the enameled brick of the Chicago Board of Trade, and peddle their wares at state fairs and in today's 24/7 hyperreal media—all in order to ensure that drilling remains pleasurable for many, enriching for the few, and unsustainable for all. The fact is that in the twenty-first century a mere ninety private and state-owned companies, in a world climbing to ten billion human actors, account for as much as 63% of the emissions that drive climate change.[12] We are thus not all equal in this mess—and that fact must be kept in sight if our politics are to matter, if we are to swing in the right direction—and if we are to understand why some of us cling to this sinking ship, while others secretly, and not so secretly, welcome the breach in the hull.

A fourth concern of this book is Enlightenment's iron cage—or to be more specific, modernity's faith (and we are talking about fossil capital's version of modernity) in the sufficiency of science as epistemology and technology as theocracy. This too is a legacy fueled by carbon—and it has, over the last two hundred years, cut across the fossil economy's many different permutations. But what we know today in hindsight is that tapping coal and steam—which closely correlated with an unabated acceleration in economic growth and technological change—had a long-lasting ideological impact beyond its obvious material impact. Because machinery and science gave us greater control over nature's energy and work, and because they corresponded with rising GDPs, they came to appear over time to the world's major players as proof of their project—as evidence of its righteousness and rectitude—even as coal and steam supplied, at the same time, those same actors with the physical engines needed to expand their empires. Today, the cornerstones of that Enlightenment project are still standing. They remain the West's preferred knowledge regime, with science and technology still operating, only partly contested, as the privileged "measure of man."[13] (Any doubters might turn, for instance, to today's focus on STEM—science, technology, engineering, and math—as exhibit A.) But that barometer broke some time ago, and its mercury leaks as toxically as the story it has told. None of us, of course, wants to throw the baby out with the bathwater, but these essays suggest that we need to put this discourse squarely back into history, into its proper place in our lives.

Finally, the essays in this book strike a different, even uncanny, register of the human experience under fossil capital. They do that by breathing life back into modernity's ontological empire, by resuscitating a repressed body—

its layers of sense, affect, emotion, touch, sentiment—that otherwise gets buried in the world's official business and all but disappears from its histo- riographical record. That is, the following essays touch down here and there on the elusive and yet bone-deep domain of infrapolitics where modern bod- ies are charged by carbon, where they feel the warmth of its heat, the chill of its absence in the blood, the eroticism of its automobility, wind in the hair, sun through the window, and the drip of its sweat deep down in the boiler rooms and back alleys of the world. Such attention to *affect*—to a feel for the world—is reinforced stylistically in these essays by a parallel challenge to the depersonalized rhetoric of academic writing. These essays challenge, at least periodically, what can be too-precious of a commitment to the fiction of detachment that while vital to our labor and profession can at times be- come a hindrance to communication, understanding, and audience. In this respect, they invite into the text a measure of play, encouraging readers to enjoy the apertures of our reason and the excess and supplementarity of life a little more than is tolerated in academic writing, so as to remind us that the language and history we have today, themselves products of modernity and appendages of fossil capitalism, are not up to the existential and onto- logical task at hand. These essays suppose, that is, that there might be some agency in contesting the disciplinary protocols that shrink our portraits of modern life down to what can feel all too often like an exhausted, dull, and disciplining affair.

And so . . .

We are not in a place to dither. The moment has reached its crisis. The chips are in, and we all find ourselves, despite our differences and similari- ties, somehow losing the long game, even if some of us are winning the short one. We are trapped, for better or worse, into using the fossil takeoff to lift us into a postcarbon future. As critic Matthew Huber writes, we have reached an impasse and are stuck with what we've got, with learning to leverage the capacities of fossil energy to generate a more sustainable and just future.[14]

No one doubts that power and privilege, economic and military might, will matter in the future, that they will determine where shrinking harvests go, which dykes get repaired, and where unpredictable fresh water supplies flow to. We know that the coming heat will not strike all of us equally. But even so none of us will be cordoned off from this species-level threat. Our children and grandchildren will increasingly sweat and worry whoever they are. Without some quick and preconceived plan for adaptation, it is quite possible, perhaps it is most likely, that we will all shift outside of the Holocene's comfort zone and watch the world's promise of democracy shake and rattle, the world's prospects for upward mobility shrink, and new feudal moats pop up around wealth and privilege.

That uncertainty is ours to share, and thus rethinking our rituals is the task of everyone. We can still hope for a future without moats, with longevity, with compassion, and with equity. But there is no invisible hand to get us there.

One

Mineral Rites

The Embodiment
of Fossil Fuels

Breathe in the warm swell of coal.
Empty your mind of chatter.
Recline into child's pose.
Let the quiet textures of petroleum caress your body.
Congratulate yourself for being here.
This is your hour, your truth.

Our erotic attachments to fossil fuels are ultra deep, to repurpose a phrase used by critics of deep-water drilling. They are rooted in minutia, in the intimate quotidian rituals of the home, workplace, streets, and stores, and such unlikely spaces as the boudoir, the lavatory, and the yoga studio. Fossil fuels compose time and space in the modern world and they leave an imprint on our emotional, aesthetic, and sensory lives. They are both interior and exterior to us. Our relationship to carbon neither begins nor ends, that is, with the pleasures of the gas-fueled automobile and the transoceanic flight, nor is it limited to the sexualized imagery of vehicular mobility and the luminance of the modern city. Such things shape the eros of modern life, especially for the world's middling and aspiring middle classes, but they are only the most transparent manifestations of a much deeper phenomenology of carbon that cuts through our lives in socially stratified ways.

So what is meant by the embodiment, or the eroticism, of fossil fuels? What might it mean to adduce a phenomenology of the present?

Fossil fuels are embodied in the sense that the modern soul is disciplined by and substantiated by, that is, given its essence within, an infrastructure of mineral energy flows (e.g., propulsive energies, ambient energies, con-

Origins. Coal in the hands of a miner. © *Siberia Video and Photo/Shutterstock*

gealed energies, etc.) that assert themselves on the flesh, the psyche, and the horizon of life. This soul, to the extent we can speak of it, Michel Foucault reminds us, is a prisoner of the body—an effect, or an instrument, of disciplinary relations that are imprinted mostly below the register of consciousness but that nonetheless lift up that body, preparing it for flight, press down on that body, burdening it with the world's heft, and endow it, albeit in unequal ways, with a good part of its potential for self-realization. Put a bit differently, the embodiment of fossil fuels describes how today's mineral rites—the body's immersion, or saturation, in the materiality and praxis of modernity's rituals—condition life and being under fossil capitalism.[1]

The Eroticism of Fossil Fuels: Part 1

From the first step into the morning shower, a saturate heat breaks through the fog of consciousness. That heat, a fossil heat, conditions the modern body to carbon. These little charges of combustion that warm up the world—and the refined chains of hydrocarbons that texture it—can appear hyperbolic and superfluous at times, and they can be acutely missed in their absence, but throughout the developed world, such carbon inputs serve as the precondition to modernity's standard of living for the world's upper and middling classes and to the sensual attachments that such a standard of living implies. The textures of the clothing we wear during the day, the food choices we make at breakfast, the tempo and tone of the evening commute, the objects we play and exercise with in childhood and adulthood, and the quality of our sleep have something to do with the extraction, the refinement, and the combustion of carbon.

We might take for illustration one hyperbolic example—the Western middle-class practice of hot yoga. Carbon's cultural work is put on spectacular display in this carbon rite that not only appropriates and repurposes East Indian culture but that appropriates and repurposes subaltern peoples and mineral ecologies around the globe. Here in the yoga studio, whether Bikram yoga, CorePower, or the local mom-and-pop variation, we have a metonym, metaphor, and exaggeration of the modern condition. Hot yoga's wild excess of steam heat, its long stretch of vinyl-wood flooring, its sheen of plate glass walls and mirrors, and its textures of spandex, polyester, and PVC mats immerse the modern body (in this case, the bourgeois body) in a festival of tactile and visual sensations that trace back to the pleasures of combusted and refined carbons. In the hot yoga studio, as in modernity's many other material rituals, the body is acclimated to the surfeit of coal's heat, conditioned to the touch of congealed oil and natural gas, and joined to a subterranean infrastructure of carbon that delivers these little sensations to the skin, ear, and eye.

It is a peculiar moment we inhabit when even the search for the self delivers us back to the coal mine, the oil well, and the boiler room.

Ambient Energy: The Body's Saturation in a Fossil Heat

To understand this structure of embodiment, we can examine its core modalities.

First, this structure of embodiment derives from the body's saturation in a surfeit of mineral heat that carries us beyond the organic forest.

This surplus heat, a fossil heat rather than a traditional organic heat, serves as the precondition of modern life. It arises from the large, controlled burn that has been going on in boiler rooms throughout the West and elsewhere for nearly two hundred years. Today we combust in fossil fuels, in the United States alone, the sustainable growth equivalent of twenty-one billion acres of forest annually, or more than double the sustainable output of the entire world's global reserves.[2] Whereas heat in the premodern world had purchase—it was often scarce, limited, hard earned, and frequently visible to the eye—heat today is simply ubiquitous, assumed, and thus invisible. It is ambient. According to philosopher Dennis Skocz we are daily conditioned to this surfeit heat that our bodies absorb unconsciously and that appears without origin: modernity's "warmth and comfort," he writes, "are not perceived as the effect of anything. They simply are experienced phenomena without a history or an anchor in anything outside themselves."[3] That heat calms the body when we step into a hot shower, generates the savoriness of soup on a kitchen stove, and provides the security of entering a warm home. It teaches us, he says, "that climate is not an issue and is controllable."[4]

The hot yoga studio makes this ambient energy visible while erasing its source. Steam rises from ventilators to provide a sensory contrast to the cool waterless air of a desert climate, and hot blasts, exceeding 104 degrees Fahrenheit, invite the body into a ritualistic sweating that is grotesque, therapeutic, and perfectly modern. In the yoga studio, this heat is elevated to totem.

Congealed Energy: The Body and Its Exosomatic Environment

Second, fossil fuels are embodied through what Victor Hugo once called the "consubstantial flexibility of a man and an edifice."[5]

The body, phenomenologists tell us, is co-penetrated by its material environment: in our case, by an architecture of congealed carbon—energy transmuted into glass, cement, and steel.[6] Today modern eyes, ears, and fingertips anticipate a material environment that differs dramatically from the world of wood and stone that the body once navigated, grated against, and enjoyed. Our bodies carry within them the rough memory of knees hitting cement, eyes lingering on the smooth curve of steel arches, and light refracted through plate glass. In this not-so-new world order, walls of steel and a terra firma of concrete are energy incarnate.[7]

The yoga studio, like the rest of our lives, takes part in this symphony, or, depending on the environment, discordance, of congealed carbons. Mirrored walls reflect the body back to itself while a sheen of sheet glass, open steel rafters, and recessed lighting provide alternately quiet and dramatic invitations to the modern eye. An aura of naturalness surrounds this house for the soul, but there is virtually nothing here in the yoga studio that does not depend on mineral heat for its fabrication.

Coal and enlightenment go hand in hand in this space.

Polymerized Energy: The Body and Carbon's Intimate Fabrics

Third, we embody carbon through the intimate fabrics we wear on our bodies and drape in our homes as accoutrements of the self.

Through polymerization—wherein monomers coaxed from hydrocarbons are strung out into chains of polymers and heated into resins—we acquire the elasticity of modern life that we know as spandex, polyester, polyvinyl chloride, and other synthetics.[8] This carbon that makes contact with the flesh, that drapes down our walls, and that is assembled like idols around our bodies, finds its way consciously and subconsciously into the poetics of daily life, as memories, nerves, and muscles become "encumbered" with these commonplace facts.[9]

This other embodiment of fossil fuels is physical and intimate in the yoga studio. It embraces the lissome stretch of spandex across the thigh, the forgiving cushion of high-density foam, and the damp touch of polyester wicking away sweat from an overheated body. The tree of life might stand boldly at the entrance to the yoga studio, giving to this space the aura of the unprocessed, but we too are polymerized here, bound in body and mind to these manufactured chains of carbon.

Propulsive Energy: The Body Served by Carbon

Finally, our embodiment of fossil fuels derives from the unspoken leisure that carbon affords the modern body as it performs disembodied labor behind our walls and in the other hidden spaces of our lives. This vast kinetic work subsidy shapes all our lives, although in different ways, depending on our social positioning.

In the modern world, mineral heat is commuted into labor—minute by minute—without remark, with each American burning sufficient carbon throughout the day to generate the labor power of nine horses, or what scholars have calculated to be eighty-nine human bodies.[10] And although that work is not always visible to the eye, this infrastructure of empowerment by carbon leaves its bold signature on the nature of our leisure, productivity, and psyche.[11]

At the close of yoga practice, I anticipate a shower that was not drawn by hand—not by my hand or any other—and I am entitled to a heat not borne of wood or delivered by foot. Electric pumps move heavy rivers of water (8.3 pounds a gallon / 25 gallons a shower) across mountains to deliver little streams to my feet; natural gas lines imbue me with kinetic energy and ready heat at the flick of a switch; and oil pipelines condition me to a universe of horsepower that emancipates me whenever I prefer not to walk, lift, pound, or swing without help.

The body might take center stage in the hot yoga studio, but that body rarely arrives on its own and never leaves without accruing to itself a heavy attachment to kinetic carbons that generate and circulate a staggeringly modern quantity of water, people, goods, resources, and heat. While we crave to leave behind the hustle of a fossil-fueled world when we enter the sacred space of the yoga studio, here too the body is moved by carbon.

Sweet *Savasana*

But for now—sweet *savasana*.

I take a final drink from a polyurethane water bottle, strip off a pair of nylon-blend Prana pants, and turn the plastic handle of this plastic basin to release the steam of a warm shower. Behind the wall, copper pipelines hiss with gas, electric water pumps churn silently, and a hot burst of water rains down in a cool evening desert. This warm cataract that gushes over my body leaves only the faintest trace of the fracking of California, the drilling of North Dakota, and the strip mining of Wyoming that gives me pleasure.

I needed this. Soaked in coal and oil, I am now fully and finally in my body.

The Eroticism of Fossil Fuels: Part 2

Breath is not universal in the modern world. It takes us back through the stratifications of power and prerogative. To breathe deliberately is a privilege. This breathing in the yoga studio is only half of the story. The fossil unconscious lurks not far below. The entropic energy of fossil capitalism issues forth.

On one side of the spectrum sits this calm breathing of the hot yoga studio where carbon is harmonized and the body tuned to its combustion. Here the potential for self-actualization and health is played out. But like every other fulfillment that carbon enables—the weekend drive, the transoceanic flight, the organic apple from Peru, or central heating in the home—a labored breathing, a spectral doubling, takes place somewhere else, whether across the tracks in the city, in a remote Appalachian strip mine, in the blight of the global industrial corridor, or in an unwritten future.

If breath is our metaphor for life, if it signals to us the body circulating in health, then this other breathing is also ours to claim. We have a moral stake in this wheezing on the other side of the tracks even if carbon's infrastructure renders it remote and other.

Ambient Energies: The Body's
Saturation in a Residue of Mineral Heat

To flesh out this other embodiment of carbon we might revisit these modalities as they appear to other bodies in other geographies in the form of entropic energy.

First, this mineral heat, which produces life and comfort in one place, produces as its alter ego an unwelcomed heat, a residue, that falls outside of the engineer's control. These eleven and a half tons of coal (or tons of coal equivalent) that are stoked into modernity's furnace annually for each American generate a residuum that expands like a gas, and literally as a gas, to affect our collective health in less salutary ways.[12] The hot yoga studio that commands that fire and brings it to bear down on us in therapeutic manner creates a certain disorder and collateral damage elsewhere.

It is a peculiar hearth that we draw our life from.

July 28, 2014. Navajo Reservation, Four Corners.

The air over the Navajo Nation is thick in some parts with the residue of burned carbon.[13] The quality of breathing here depends on where you stand. But, as of today, the Environmental Protection Agency (EPA) has decided to enforce emissions regulations that might help to protect the lungs of local workers and neighbors (from mercury, sulfur dioxide, and nitrogen oxide) and to reduce the smog that hangs over the Grand Canyon.[14]

This sovereign nation in the heart of the American Southwest is home to the third and fourth most polluting coal plants in the region, the Navajo Generating Station and the Four Corners Power Plant, and it is surrounded by five other massive coal plants privately or publicly owned, each sitting, like opportunists, just on the edges of the reservation.[15] The reservation's Kayenta Mine (an open pit mine in the Black Mesa region owned by the global giant Peabody Energy, recently infamous for its coal slurry pipeline) and the nearby El Segundo and San Juan Mines (also large privately owned strip mines) have come to define the landscape of this region just as surely as its open ranges and rain-carved mesas. Coal mining here means that more than four hundred acres of reservation land are stripped clean each year of their original ecology. And while coal mining and power generation leave their visible footprint on reservation life, there is also the invisible mineral legacy of five hundred abandoned uranium mines that continue to seep radioactive gasses into the lungs of nearby Diné men and women.[16] In the words of the Navajo Nation's Department of Justice, the reservation functions as an "energy colony" of the modern United States and corporate capital, a place where the externalities of a mass market in cheap heat and electricity are off-loaded into the lungs and bodies of the nation's less privileged citizens.[17] Not-in-my-back-yard politics for those of us living in the Sunbelt means in the backyard of the Diné.

The Navajo reservation, as well as its immediate environs, has long served as an epicenter of Southwestern energy production. The reservation's economy has been dominated since midcentury by coal and uranium mining (making up 50% of the tribe's receipts), and the electric-generating plants that sit on or just off the reservation have played a critical role since that time in the rise of the region's high-tech sector and the culture of the Sunbelt that has grown up with it, as historian Andrew Needham has argued. In fact, the scope of this busy energy intersection's influence is remarkable. Until divestment in 2014, even the Los Angeles Department of Water and Power, located 546 miles away across harsh deserts and mountains, had a firm hand in the ownership of electricity generated here.[18] Which is to say that despite the reservation's seeming remoteness, the Diné people's lives are bound up in a material infrastructure of bourgeois comfort that stretches back into the Black Mesa mines of northern Arizona, into eastern Kentucky's rich mountain ecologies, and into various other subaltern ecologies around the world.

Fossil capital expresses itself differently here than it does in the hot yoga studio. The poverty rate is stuck at a stunning 42% for families who live on the reservation, despite the profitable quantity of exports of coal and electricity delivered offsite. Thirty percent of Diné homes (or a figure that is worse than the impoverished Himalayan nation of Nepal) lack running water and electricity, despite the fact that locals look up at high-power transmission lines that move such resources to middle- and working-class families off the reservation.[19] Moreover, the material dialectic and cultural logic that connects the hot yoga studio to reservation life is evident in the patterns of environmental health on the reservation. Although air quality on most of

Navajo land typically falls within acceptable limits because of its vast size and high winds, the air gets thicker—and filled with pulmonary pollutants like nitrous oxides, sulfur dioxides, and, of course, radon—for families that live near strip mines, uranium mines, and these two-thousand-megawatt power plants that have a poor history of emissions and waste control. Aerial contamination in this high desert is, in fact, so palpable that Grand Canyon tourists complain that the reservation's smog is ruining their family pictures and astronauts once claimed, back in the 1960s, that they could see from outer space a plume of emissions rising from the reservation's Four Corner's Power Plant. Perhaps worse for local lungs is the stark contrast between the clean air of the hot yoga studio and the indoor air pollution that affects a third of the reservation. Although official World Bank statistics claim that the United States is now 100% electrified, 30% of Diné homes lack electricity (and running water), and thus many engage in the dirty business of combusting wood or coal inside homes that leave respiratory pollutants deposited onsite in Diné families' lungs rather than offsite in someone else's.[20] Breathing on the reservation can thus be hard in metaphorical, and sometimes literal, terms.

The Navajo reservation cannot be written off as an exception to modernity. It is, as Needham tells us, among the most modern of geographies in the United States.[21] It is the other half of a dialectic of therapy and injury generated by a fossil heat that draws systemic suffering on the reservation into close quarters with the deep breathing of the modern yoga studio.

Congealed Energies: The Embodiment of Carbon's Discordant Environment

Second, this "energy deepening" that produces the exosomatic environment of modern life—symbolized by the mirrored walls and clean steel lines of the hot yoga studio—also produces as its alter ego a shadow world of hard-breathing, wheezing bodies, of inhospitable concrete, and of contaminated industrialized ecologies across fossil capital's other sacrifice zones.[22] Once that wheezing was heard nearby in Sheffield and Pittsburgh, where emissions from heavy metal, cement, and glass production were part of a working-class life that included heightened cardiovascular diseases, cardiopulmonary diseases, lung cancer, brain cancer, ovarian cancer, and a host of other long-term health risks from heavy aerial contaminants, but today that same wheezing can be heard along a shifting geography of precarity that binds the health of the world's peoples to one another across state lines, continents, and oceans.

December 21, 2015. Hebei Province.

Beijing is on red alert for the second time this month. This municipality, which is surrounded by the industrial province of Hebei, home to a good chunk of the world's steel production, is four days into a particularly harsh winter inversion that has, like a tight lid on a pot, trapped carbon emissions over the city. The air pollutant PM2.5, particulate matter thinner than a string of hair, circulates here at seven times the recommended maximum exposure. That toxic dust gets sucked into the lungs of industry workers and local residents, causing a host of respiratory problems, especially for the very young and old. The air here is, residents claim, "like a toxic gas," and it can hurt to talk after long-term exposure.[23] Beijing's municipal government has consequently shut down 2,100 factories, confiscated personal barbecues, and warned the city's residents to stay inside near air filters unless it is absolutely necessary that they leave their homes.[24] What the quality of breathing will be tomorrow in this city depends largely on what the wind chooses to do.

Hebei Province is among the world's dirtiest provinces in terms of air quality, and it is home to modernity's heavy industries, including steel, glass, and cement. China produces today nearly 50% of the world's steel (in addition to 50% of its cement and 20% of its glass exports), and Hebei province plays a very large part in that manufacturing.[25] The air here is thus not really a local problem, or at least not only a local problem. It is a global one, as under late capitalism, the world's smokestacks have been reassigned spatially so as to transfer risk and wealth with an almost Platonic perfection.

In contrast to practitioners in the hot yoga studio, where the point is to breathe deliberately, to take in the environment and lose something of oneself to the world, those living in the global industrial corridor find it healthier to put up a protective barrier against their environment. The images, in fact, are stunning. Across Beijing and Hebei Province, children and parents, laborers and office workers, are found in the streets, in hallways, and on buses wearing air-filtering masks, geared up for what looks like, and is in fact, a chemical attack. Residents here band together not only to pray and meditate but to purchase indoor air filters to strain out particulate matter that they know is the corporeal cost of progress. If 4.2 million people died prematurely in 2016 from ambient outdoor air pollution around the world, that burden is not equally shared.[26] Bodies that congregate around the world's mines and heavy industries are sacrificed to steel, glass, and cement production; the costs to the unborn and to the young include reduced cognitive development, an increased rate of autism, smaller head circumference, and elevated anxiety and depression in addition to the expected increases in cancers and pulmonary diseases.[27] A nineteenth-century smog still hangs over this twenty-first-century life.

Beijing, Suzhou, New Delhi, and frequently in places surprisingly close to home—there is systemic injury built into the world's walls of steel and plate glass.[28] These sobering spaces that are chained to the yoga studio remind us that we have never left behind the age of steel and concrete and that a savage industrialism has only been ramped up and relocated to someone else's ecology. But we had, until recently, breathed a little easier thinking things had changed.

Polymerized Energies: The Erotic Residue of Carbon's Textures

Third, the tactile pleasures we get from carbon's intimate objects—the stretch of spandex and Lycra and the soft cushion of a polyvinyl chloride yoga mat—signify the athleticism, freedom, and health of the mineral moment, but this lissome stretching, this signature agility of the modern, also draws us closer to the fracking field and to the world's petrochemical corridor.[29] The plasticity we gain from living among foam rubbers, synthetic floorings, polyester textiles, and olefin carpets ties us, that is, to a barricaded, restricted, and off-limits world of hydrocarbon refineries where disability and anxiety are the counterpoint to the health of the yoga studio.

August 1, 2014. Kaohsiung, Taiwan.

Today, a petrochemical pipeline exploded in the city of Kaohsiung, killing 28 people and injuring 268. The explosion left a 6.5-foot trench cutting across 3.7 miles of city streets. It also left 23,600 households without gas supplies and more than half of those families without running water. According to residents, the pipeline explosion "sounded like a bomb," and one older resident initially thought that mainland China had decide to shell the city. In the days leading up to the explosion residents had been complaining that an unseen but noxious gas, propylene, a chemical ingredient derived from the catalytic cracking of natural gas and used as a feedstock for food additives, drugs, and, among other things, the synthetic straps and blocks of the hot yoga studio, had been leaking fumes from one of the city's buried pipelines. That gas turned out to be explosive when it hit the sewer system, and the resulting violence brought the underground geography of petrochemical refining a little closer to the quiet reflection of the polymerized yoga studio.[30]

Kaohsiung City is the core industrial zone of Taiwan, and Taiwan has since the 1980s pivoted on the petrochemical industry.[31] Plastics, chemicals, and oil are three of the island's top ten exports, and hydrocarbon cracking, the process of turning oil and natural gas into chemical compounds and chains of polymers for synthetic fibers, mats, carpets, and pipes, is a busy business not only here but across the global chemical corridor.[32] Kaohsiung City, like the rest of the world's petrochemical geographies, has consequently been what environmental scientists call a "hot spot" for its toxic discharges into water, soil, and air.[33] For instance, recent inspections at the now-defunct Renwu polyvinyl chloride facility of the world's second largest PVC manufacturer, the Formosa Plastics Group, revealed that the groundwater and soil contain exceptionally high levels of carcinogens from plastic production like benzene, chloroform, vinyl chloride, and dichloromethane along with unpronounceable ones like 1,1,2-trichloroethane, 1,1-dichloroethylene, tetrachloroethylene, and trichloroethylene.[34] Further downstream Taiwanese bodies host wildly high loads of the endocrine disrupter BPA, or bisphenol A, from the waste stream of another plastics manufacturing plant in the Linyuan industrial zone.[35] These wastes, like most other petrochemical releases here and elsewhere, are part and parcel of our synthetic world, and they are closely associated with a range of disorders, including nervous system dysfunction, liver cancer, kidney and heart failure, and reproductive problems. But there is nothing unique to Kaohsiung or Taiwan's largest petrochemical producer. BASF, Dow Chemical, ExxonMobil Chemical, Illiopolis, Illinois, Point Comfort, Texas—the story is nearly the same.[36]

Modernity's elasticity has its trademarks. Made in Taiwan, made in Bangladesh, and made in the USA. These little signifiers on our water bottles and quick-dry towels tie us into a global ecology of hydrocarbon refining that offloads injury into other bodies and other soils. Ethylene vinyl acetate foam blocks, Lycra yoga pants, plastic water bottles, polyvinyl chloride mats—the things that make the yoga studio a space of emancipation link health to a panoply of unfamiliar petrochemical compounds that stack up as among the world's most important and hazardous trade goods. Today, this polymerized world is so universal that it sheds a residue of microplastics day and night (from laundering beads to synthetic fabrics like yoga towels and pants) in such high quantities that scientists find emancipated plastic fibers in every water sample on the planet.[37]

The external costs of today's flexibility are not simply an error of faulty regulation; they are not simply a moral error in our ability to care about other people: these little doses of toxic waste are a founding part of economic growth under the terms of fossil capital. They are part of a mineral moment that makes wheezing and worrying about the lungs of children as modern as the calm deep breath of meditation in the yoga studio.

Propulsive Energies: The Laboring Body under Carbon

Fourth, the movement we derive from combustion, modernity's labor subsidy, is also a premise of the hot yoga studio, even if the human body, rather than the mechanical one, is on display here. This hour of quiet meditation in the middle of the day is made possible for North Americans by a hidden endowment of labor that is equivalent to dozens of human bodies working day and night. This nearly invisible labor force we depend on derives from charges of combustion that allow bodies to circulate without exertion in three-thousand-pound Priuses, water to be pumped interstate across mountains and deserts to bathe the practicing body in humidity, and plate glass walls and steel beams, foam blocks and elastic fabrics, to be shipped across oceans to generate the mood for quiet self-reflection. The automobile, the container ship, and the water pump—the road, the flight path, and the shipping lane—these are embedded in the hot yoga studio, making the textures, the placement, the aesthetics, and the leisure of this space possible, even if for each mile of the road, for each lift of the escalator, and for each pump of the aqueduct that work demands a little sacrifice from some other remote environment and from the people who live in it.

September 11, 2005. Lake Charles, Louisiana.

Residents living near the Pelican Refining Company in this state got a strong inhalation of air laced with BTEX (benzene, toluene, ethylbenzene, and xylene) and hydrogen sulfide today. At low concentrations, the smell of the hydrogen sulfide would have stunk like rotten eggs. At higher concentrations that same gas would have entered the lungs, numbed the nervous system, and rendered itself undetectable to the recipient. Breathing downwind from this refinery, one of the Gulf Coast's asphalt manufacturing facilities, meant nausea and headaches for local residents, and over the long term, it meant the body's chronic exposure to abnormally high emissions of the unwelcome volatile organic compounds familiar to oil refining, including the cancerous quartet, benzene, toluene, ethylbenzene, and xylene.[38] Mechanical labor and its product, carbon mobility, require processing the world's oil and natural gas, and that can be a hot, leaky, and unhealthy business.

Four of the world's largest oil refineries sit near here as part of a petroleum and natural gas infrastructure that runs for four hundred miles along the northern Gulf of Mexico from Baton Rouge, Louisiana, to Houston, Texas. Not surprisingly, the most contaminated zones are peopled predominately by low-income African Americans and low-income whites. It is here that, among other things, sour crude (with its high impurities of sulfur) from the Gulf, from Mexico, from Venezuela, and from Alberta's tar sands is converted into gasoline, diesel fuel, jet fuel, naphtha, and a host of other fuels and chemical products. The United States remains today the "refiner to the world" and this region is at the center of it all.[39] American asphalt—the key to the road—is one of the many products that comes out of this leaky corridor where nearly 50% of the petroleum that reaches the United States is refined.[40]

As the anthropologist Michael Watts explains, there is a chronic leakage, a systemic atmospheric, aquatic, and terrestrial contamination, that is integral to modern work and mobility.[41] Most of that leakage goes undocumented, but periodically it comes to the surface. What is remarkable in the example at hand is that in the twenty-first century, in the presumably developed world, in the world's third largest producer of petroleum, neighborhoods breathed deeply in for years as a refining company, answering to dictates coming from the nation's financial energy center, Houston, operated with no environmental budget, no environmental regulation plan, and no employee tasked with conforming to state and federal regulations. The costs went undocumented for years. They included unwelcome emissions whenever sour crude was off-loaded at cargo docks, further seepage into the ambient air whenever that oil was stored in tanks without an operating roofing system, and chronic leakage into neighborhoods whenever the refinery operated without appropriate scrubbers or flares. It would be comical, if it weren't depressing, that the only plan in Lake Charles was to doctor the books and to direct endangered employees to trot off to Walmart to purchase flare guns to shoot at the facility's smokestacks whenever the plant's faulty flares went out.[42]

The air around the Lake Charles refinery complex is just a small symbol of the chronic leakage, legal and illegal, that goes into the air, soil, and water from oil and gas processing across the world's energy frontier. That leakage (symbolized more grandiosely by the Macondo blowout that affected as much as sixty-two thousand square miles of this region's gulf coast) occurs when fossil fuels are pulled out of the ground, whether in Inez, Kentucky, in the Niger Delta, or in the forests of Amazonia; it occurs when oil is transported and refined, whether by ship, rail, or truck, and it finds its personal expression in the CO_2 that pulses out of the automobile on the way to the yoga studio, heating up the planet just a little more each mile.

If the hot yoga studio defines itself as a place of repair and healing, this shadow world is a place of injury—a modernized "wasteland," to quote one critic, where subaltern peoples engage in "toxic marches" for the sake of their children and where the oily footprint of one state includes as much as sixteen thousand pounds of petroleum waste products annually for each person, with that waste dumped disproportionately into communities of high unemployment, high illiteracy, and ill health.[43] Carbon's mineral rites emancipate the privileged body in one place, but they degrade these subaltern bodies elsewhere.

Savasana: Corpse Pose

But life persists. And so to end, a final belly breath.

We relax a second time into *savasana*, or corpse pose, this ritual of death in life, where we register the entanglements between breathing and cessation and pause somewhere between activity and stasis, health and injury. As we subside this final time into oblivion, unwinding through the tensions of the day, we have the opportunity to give up the modern self, to wake up renewed and reborn with a new awareness and with fewer attachments to the material world in hand. We take together a final collective breath and bow to the god that is within us.

Namaste.

Two

Carbon's Social History

A Chunk of Coal from
the 1912 RMS *Titanic*

Fossil fuels have a social life.[1] Coal appears static. Jet black. No more than surface. Natural gas shows up unseen. An emptiness inside of a conduit. A sound that can't be held. And petroleum comes as movement. Liquidity. A transition that resists settling into a single state. But this opacity, this impenetrability, the ungraspable essence of carbon, belies a simple fact.

Fossil fuels are not flat. They are not merely surface or opaque. Like still waters, they run deep, and can be plumbed. To the human eye, the coal heap is all we see. But this heap is something more. It is a conurbation of history, an aggregation of small pieces of the past that have been lumped and jumbled together. Each shovelful of coal, each tankful of gas, each pulse of the ventilator carries us backward into this history, taking us through commodity chains that cross time and space and that put us into touch with people in Appalachian valleys, the deep ecology of Silesia, the pits of the Powder River Basin, and an unwritten future of drought and baked soil. Carbon's history might be cast aside as a geological matter, but it is historical and pointedly human.

The humanities—the fields that deal with people's memories, fears, injustices, values, and passions—were slow to take an interest in fossil fuels, as our attention had been turned primarily to other important social concerns rooted in race, class, gender, and ethnicity. But we have taken the cue from geologists and engineers to ask questions about the life cycle of carbon and about how these energies enter into, shape, and inform our social lives, our economic beings, and our thought patterns in deep entanglements. To critic Imre Szeman, knowing our energy dependencies is vital not only because they prop up the materiality of everyday praxis but because, as he explains, they "animate and enable all manner of abstract categories, including freedom, mobility, growth, entrepreneurship, and the future."[2]

That is to say carbon is not simply "fuel," as we understand the term. Not merely the chemical matter that gives to living beings their flesh and bone. It is, and has been, the bioenergetic basis for *Homo sapiens* flourishing since the time we first broke with nature's more humble solar cycles in the nineteenth century.[3] It is the input that makes us modern, the root of our coming crisis.

To understand who we are today means understanding energy flows and energy infrastructure. It means coming to terms with today's concealed, and increasingly corroded, substructure of pipelines, mile-long coal trains, gas-fired power plants, and shoreline refineries that lift up the modern subject and distribute well-being and disability in the Anthropocene. What is identified as 03266 can serve us as an example of what it means to see ourselves as fleshed out by, or subjected within, this infrastructure of carbon.

A Chunk of Coal from the RMS *Titanic*

My chunk of coal comes with a registration number, number 03266/25,000. It arrives by mail in a commemorative gift box accompanied by a "certificate of authenticity" guaranteed by the RMS *Titanic* Corporation, the only entity that has the legal right to forage goods from that sunken steam liner. Packaged and sold in commemoration of the ship's one-hundredth anniversary, this piece of coal is a limited edition that goes for around thirty dollars. If fossil fuels are mostly anonymous, abstracted like wheat, rice, and beef into a handful of predetermined grades for the market and for end-use consumption, my coal is meant to be individuated, to refer us back to a story: it has a personality and provenance not typically accorded a mineral fuel.

03266 once rested inside one of the *Titanic*'s thirty-four coal bunkers. Its latent energy, a tiny flame that would have fueled a light bulb for a few minutes, was meant to be aggregated each day with another 850 tons of best Welsh coal, a form of subanthracite coal that burns at a clean two thousand degrees Fahrenheit and that was the fuel of choice for energy-hungry steamships before the shift to petroleum. The *Titanic* sucked up such coal voraciously. With 159 furnaces and seventy-seven stokers and trimmers shoveling in coal at any given time of the day and night, it could consume thirty-five

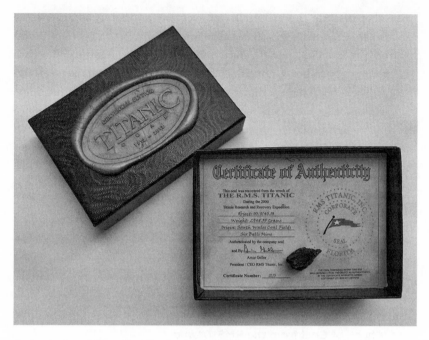

A chunk of coal from the *Titanic. Reproduction from eBay*

tons of the stuff an hour.[4] When that energy reached a pitched heat, it generated an astonishing torque, a mechanical pull equal to half a million men rowing their hearts out (i.e., 46,000 horsepower units, or 460,000 manpower units).[5] While the *Titanic's* paying passengers would have been mostly oblivious to this coal that burned just below deck as they sat to plum pudding in the third-class cabins or played squash in one of the first-class amenities, that was not so down below. Laborers deep in the ship's hull would have known this coal intimately. Members of the black gang—trimmers, greasers, and coal stokers, the latter of whom were called firemen—would have experienced it firsthand as it sunk into the capillaries of their lungs, the pores of their necks, and the grooves of their hands.

Coal had a social topography to it onboard the *Titanic*, and it reproduced a certain type of social order.

The work that my piece of coal would have performed on the *Titanic* was, however, not straightforward. Energy is protean, and coal can be commuted into different states (and put to different tasks) before it dissipates in a cloud of carbon dioxide and trace elements like mercury. On the *Titanic*, the

Fossil capital at sea. The *Titanic* and *Olympic,* 1912. *Robert John Welch/ Wikimedia Commons*

bulk of coal's business was to provide the ship's propulsion—to turn its two-story-high steel propellers at seventy revolutions a minute. But propulsion, propulsive energy, was only its most obvious function. After being pitched into the ship's double-end boilers or front-loading Scotch boilers, this piece of coal would also have passed in the form of steam through a low-pressure turbine and charged up the *Titanic*'s four-hundred-kilowatt electric genera-tors, which released the capacity of a small city's power plant (while also supplying a little extra oomph to the ship's propellers). From that point on, converted into electricity, my piece of coal would have circulated through two hundred miles of electric wires as its purpose shifted from brute force to the more delicate tasks of lighting, heating, and running motors to gen-erate the many pleasures the *Titanic*'s passengers had paid top dollar for in anticipation of eating, drinking, conversing, and making love at sea.[6]

The ship's parent company, the White Star Line, had, in fact, not adver-tised force or speed as its main selling point. That was the Cunard Line's claim. Rather it sold modern comfort at sea, ambient energy—electric chandeliers glinting from first-class smoking rooms, ventilated heat pulsing through bedroom chambers, and therapeutic steam soaked up by bodies lounging in the ship's Turkish steam baths.

Coal was not merely forceful on the *Titanic*. It was exquisite.

The *Titanic* was, and is, the Western world in miniature, a microcosm of the fossil order on land, the premier achievement of fossil capital. And as such, we can turn to it as an example of the high-energy and stratified social order that carbon fuels. In that world, my chunk of coal was foundational, and so we might ask whose lives was it meant to touch, in what ways, and what type of social power it was intended to reproduce.

Embittered Origins

We know certain things. The RMS *Titanic* began its maiden voyage out of Southampton, England, on April 10, 1912. That meant it left in the midst of uncertainty, exactly four days after the end of a month-long coal strike. That national strike had seen nearly a million coal miners walk off the job in protest of a lack of a minimum wage for their work. Certain high-grade coal did not arrive in the Southampton shipyards for the *Titanic*'s launch, in other words, because miners in the nation's collieries had decided that they were being exploited and that they might collectively do something to address their insecure standing under fossil capital. The *Titanic*'s departure thus took shape within the familiar cauldron of working-class resentment and bourgeois privilege that defines the world's energy chains.

My piece of coal purports to come from the Six Bells Mine, a mine in Abertillery, South Wales, originally sunk back in 1863. I have my doubts about that, but let us play along. Let us take that as an origin.

The Six Bells Mine, or the Arael Griffin Colliery (as it was also called at the time), was one of 562 mines making up the coal industry in South Wales, a titanic industry in itself that, on the eve of the ship's launch, employed over two hundred thousand men and had set in motion a massive internal migration southward toward the mines. The fossil economy was no small matter in this part of Wales. It supplied one-fifth of Great Britain's fuel needs and as much as one-third of the world's coal exports, with its reach extending to the Continental powers of France and Italy, central and eastern Europe, colonial Egypt, and emerging economies across the Atlantic in Brazil and Argentina.[7] Consequently, when the mines of this corner of Wales shut down, they shook the energy infrastructure of the world's workshops and threatened to make global capitalism's machinery cough and sputter.

03266 traces back to what were deep social contradictions in the emergent fossil economy. It is the mineral trace of a commodity chain that once bound the lives of coal miners on the world's mineral frontier—in networks that were socially invisible but materially tangible—to its first-class passengers sitting back, many miles away, to a cigarette in an elite smoking room. By traveling down coal's commodity chain we can bring back into the *Titanic's* history, and into our own lives, a branch in modernity's genealogy that otherwise gets left out of the family portrait.

The *Titanic* went nowhere without coal. It depended for its thrust on the type of high-grade carbon that could be found in South Wales and in a few other prime locations around the world where geological conditions had been just right to create hard coal with few impurities. As with most of the world's mineral frontiers, the rise of the mining industry in this part of the country had turned it into one of modernity's toughest geographies. Today, people talk about "tough oil" to signify the high costs we accept and the far ends to which we go to drill oil, but there was never anything other than "tough coal."[8]

The Six Bells Mine—where my piece of coal comes from—was, for instance, a place of economic morbidity, of inadequate wages, high injury rates, poor bathing and housing facilities, insecure welfare, and class bitterness—a sharp contrast to the plush comforts of the *Titanic*.[9] While work in the mines provided, of course, much-needed jobs in the bare-knuckled economy of high industrialization, it never managed to provide a secure living, or many creature comforts, no matter how dedicated or frugal a miner was. The Protestant work ethic, the mythology of modernity's middling classes, made little sense here amid the coal fields' deep structural inequalities.

The origins of this energy commodity chain begin, in other words, with class risk and insecurity. Injury alone gives us some indication of the risks and stressors faced by mining families at Six Bells. In 1899, 1902, 1903, 1904, 1905, 1906, and 1909, local papers report that young miners in the colliery were crushed to death from falling slate in the mine, leaving children and wives behind, to the prospect of poverty.[10] Similarly, in 1907, 1908, and again in 1910, miners in this pit were severely disabled or killed in other types of mishaps, such as run-away trams and steel chains snapping.[11] Injury was, however, only part of the precarity that followed this piece of coal. So too was financial hardship, and its concomitant, class struggle. Workers at the Arael Griffin Colliery struggled to make ends meet and to provide something for their children, and they consequently sought, like other miners in the region, to leverage the modicum of power they possessed to unionize.[12] This was, after all, a subaltern space deep inside the belly of the modern world where people still spoke using phrases like "masters and men" to signify the near-Platonic stratification of the economy's class structure.[13]

The fossil unconscious has always taken its richest material from such deranged workscapes.

To get a glimpse of it at Six Bells, we need only look at contemporaneous events at the Cambrian Combine, a stone's throw away in the Rhondda Valley. Here the simmering resentments of miners at the start of this commodity chain found public expression in what would turn out to be a dress rehearsal for the coming national strike that threatened the *Titanic*'s launch.

Havoc. Looting. Terror. Such runs the description of a wild night of class rioting in the town of Tonypandy.

On that famous night, at a time when the *Titanic* was still under construction in Belfast, frustrated coal miners vented their resentments against fossil capitalism by attacking mine property, busting up shop windows, looting stores, and harassing police in a loud outburst of class anger that shocked the sensibilities of the town's bourgeoisie. The catalyst appears to have been a decision made by the owners of the Cambrian Combine, a recently formed mining cartel that controlled the employment of twelve thousand men in the Rhondda district, to open up a seam in one of their pits that was particularly hard to mine. Although miners at the pit did not complain about the hard labor required to cut coal from the new seam, they did complain about the pace and quality of that work because it went very slowly and produced a lot of dead rock. Since employers only paid for coal carried out of the mine and not for the overburden, the cost of working in hard places like this one was a deep cut into miners' wages. The issue boiled over in September of that year when eighty miners refused to go back into the pit for the price being offered and when management made a lesson out of them by locking them out—and then, to emphasize the point, locking out all other eight hundred miners working elsewhere in the pit. That act of class discipline provoked a wildcat strike that lasted for ten months and that closed down all of the mines across the Cambrian Combine. The disruption was worrying enough that the government sent in federal troops.[14]

What makes the Tonypandy riots significant is that they illustrate the repressed class hostility that sits right behind the fossil order. They exemplify working-class resentment against the structural violence that makes modernity's mobilities and emancipations possible. In this respect, they give us a rare public view of the underlying psychic gulf between the world's upper class and its other carbon consumers, sitting on a ship like the *Titanic*, and its producers sweating it out on the mineral frontier. A riot like that at Tonypandy, in which mothers and wives collected rocks for their sons and husbands to throw at property, comments on the unstable home that we have made together. We see in it a momentary class catharsis aimed at the false appearance of progress that the mineral order posed—a collective "euphoria of release" from the inequality and pent-up paralysis lying at the bottom of this energy infrastructure. Here the core elements of fossil capital—property, law, order, mineral rights—signified something more oppressive to miners than they did to the world's first-class passengers.[15]

As for my piece of coal: it was caught up in this topsy-turvy world of false appearances, in an age of electricity and steel that spoke confidently of progress but sublimated its externalities, both its human and ecological casualties. The Arael Griffin Colliery played only a small part in that world, a metonym for the larger condition. And thus like most of the world's mines, it flew under the radar, absorbed into the industry's generic history. But one only needs to wait long enough to see the destructive contradictions in the fossil order surface wherever it lives. For Six Bells, that moment came fifty years later, after the *Titanic* had been resting on the seafloor for decades. One fateful day in 1960 the mine exploded from high concentrations of methane and coal dust in a national tragedy that delivered forty miners to a deep and early grave.[16]

To this symbolic and literal disfigurement we trace my piece of coal.

We have gotten ahead of ourselves, however. Let us take a step back.

03266 is not perfectly transparent about its origins. We can't be certain, without conducting a chemical sampling of its content, that my piece of coal came out of the Arael Griffin Colliery. Relics and memorabilia like those sold on eBay traffic in a marketplace of questionable authenticity. They have contested provenance. And so it is quite possible my chunk of coal came from somewhere else.

Coal was a problem on April 10, 1912. The *Titanic*'s launch was, in fact, in some doubt, because, despite the end of the national coal strike just a few days before, the gap in the movement of coal from mine mouth to seaports had not yet closed in places like Southampton. Collieries, railways, and dock loaders—these key nodes in this energy infrastructure—were still out of sync. What that meant is that high-grade coal from places like South Wales and Newcastle was in short supply and the *Titanic*'s commissaries, who on any other day would have turned to the Lewis Merthyr Navigation Steam Coal Company for their coal, had to look elsewhere to fill the ship's massive bunkers.[17]

To get that job done required the *Titanic*'s parent company to flex its financial muscle. Economic and social power determine, after all, where energy flows to, and the *Titanic* did not lack in those departments. In addition to being a physical titan, it was a financial juggernaut, one of two crown jewels in a recently capitalized holding company, the International Mercantile Marine. That holding company was the centerpiece of a transatlantic cartel that had combined the White Star Line, the Leyland Line, and the Dominion Line into one entity while cutting deals with its main German competitor and assuming part ownership, under a different entity, of its Dutch competition. This private trust—capitalized at a $170 million investment (or approximately $4 billion in today's currency) was a pet project of some of carbon's biggest powerbrokers, and it thus tied the ship's voyage back to global players like the banker J. P. Morgan, London investors, and Standard Oil pipeline interests.[18] These were people—once called robber barons—who yanked commodity chains, threw their weight around to restructure entire industries, and tugged on people's lives whenever they acted. Thus behind my piece of coal stood the cartelization of shipping and a financier elite that gave the *Titanic* the social power to requisition scarce coal from elsewhere.

There is no certificate to tell us where this other coal came from. But we do know that it was taken from at least two different sources. The first source derived from inbound ships coming from the United States. These included the *Oceanic* and the *Majestic*, each of which had bunkered extra coal while in port on the East Coast of the United States. That coal purchased on the other side of the Atlantic tied the *Titanic* into a global energy infrastructure that spanned at least two continents and that thus pit the interests of coal miners in Pennsylvania and West Virginia unwittingly against those striking miners in South Wales. A second source came from other International Mercantile Marine ships harbored in Southampton. The *Titanic*'s owners had instructed the ship's commissaries to remove coal from the company's lower-priority voyages, to "cannibalize" its other ships, as one writer puts it, so as to ensure that the *Titanic* ran on time.[19] These ships included the *St. Louis, New York,* and *Philadelphia.* And although we don't know, for sure, where the coal in their bunkers originated, we can presume that it came from one of the high-grade fields in Appalachia, South Wales, Newcastle, or the Scottish Black Country, thus tying the *Titanic*'s launch to a small constellation of high-grade coal mines on the transatlantic mineral frontier.[20]

The specific origins of 03226 are mostly immaterial in this case. Inequality and disarray in the coal fields were then—as they are today—baked in, and that was especially true in the early twentieth-century coal fields. Whether we follow yesterday's coal freightage from Southampton back to Six Bells, today's high-powered transmission lines from Phoenix to the Navajo Reservation, or our natural gas pipelines from San Francisco back into Wyoming's fracking fields, we end up at the origins of a profound cultural disorder—a dizzying moral discombobulation and topsy-turvy state of affairs that is written into our economy's constitution. South Wales was different from Appalachia, Alabama from Newcastle, and the Powder River Basin from today's Inner Mongolia—but these are all demented spaces where nature and society have been turned on their heads and where the volatility of the fossil economy belies the images splayed on modernity's poster boards selling that economy back to us as freedom and security.

But let us return to ship.

When the *Titanic* set out on April 10, 1912, an unplanned coal fire smoldered in stokehold 10. Fuel is volatile, and when coal gets exposed to air it oxidizes and releases methane that can catch fire at higher concentrations. This unplanned coal fire in the ship's bunkers had started before the *Titanic* arrived at port in Southampton, and it was still burning on the day that the *Titanic*'s 2,228 passengers embarked. Port authorities knew about it, but they gave the ship its certificate of seaworthiness anyway. But for the next three days, enclosed in walls of steel, with passengers dining upstairs, eight to ten coal stokers worked tirelessly in stokehold 10, emptying coal out a 365-ton coal bunker—shovelful by shovelful, hosing down its burning embers and trying to extinguish this unwanted burning. It took three days, until April 13, for that fire to be put out.[21] But this unplanned waste, this element of nonhuman agency, this brutal laboring below can represent for us the systemic social and material chaos that propelled the *Titanic* and its world forward.

Fossilized Performances

The genealogy of 03266 leads back to the collieries. But it also points forward to the boiler room, and from there, to the *Titanic*'s many pleasured sites of consumption. My piece of coal was not only meant for propulsion. It was also meant to do cultural work by propping up certain class, ethnic, and gender performances on board ship. In fact, its social life, to the extent that we can speak of it in this way, was meant to come to maturity in the service of a high-energy life that parceled out privilege and labor along class lines and carefully policed spatial boundaries.

To start, we can follow coal down into the boiler room—to the stokehold where combustion meant lower-class men sweating life out for carbon consumers upstairs. There was no mixing of unrefined coal and the pleasures of consumption. This part of the ship was architecturally off limits to the *Titanic*'s passengers, just as its laborers, known as the black gang (because of the coal that clung to their faces, clothes, and lungs), were also located out of sight. In fact, both the ship's crew and its passengers moved along carefully scripted pathways, intersecting, when they did, in predetermined ways. No one with the exception of stokers, trimmers, and engineers entered, for instance, into the raw work space of the boiler room where my piece of coal once sat: that required special dispensation from the captain. But for the "black crew," the ship's underbelly was a perfectly familiar space.[22] These hardened men moved in a closed circuit that ran from the firemen's quarters and its mess hall forward of the third-class cabins, down steel staircases to a special firemen's passage, and ultimately into the ship's noisy bunkers and boiler rooms.[23] The symbolism of the ship's social architecture had, in fact, a Marxian clarity to it: low-wage labor, blackened by coal dust, passed daily under the feet of the ship's upper-class passengers, who were quite literally relaxing only a few feet above them in oriental steam baths fed by the coal (and the lives of these other men) below. It is hard to imagine a clearer metaphor for being structurally underfoot in the fossil economy.

The stokehold might have been materially central to the operations of fossil capital but it had none of the *Titanic*'s sumptuousness. The deep hull of the ship was an austere space of congealed energy, of thick steel, characterized by the sounds of metal scraping, the texture of coal crunching, and the roar of the furnace's heat. Without rugs underfoot, without drapery to soften its hard metallic edges, and without fresh breezes to clear out its oppressive air, the ship's boiler room reflected modernity's savagery.

The work performed in the stokehold had a dose of cruelty to it. Heavy sweating was done to the sonic boom and metronymic pace of industrialization. For the *Titanic* to run at about twenty-four knots an hour, or seventy to seventy-five revolutions per minute, to keep its needle "in the blood," as it were, and for it to build the head of steam pressure needed to supply the ship's electric generators, every four-hour shift, night and day, required fifty-four coal stokers, twenty-four trimmers, and five lead firemen tending to multiple furnaces simultaneously.[24] That labor was not casual. It was done at a quick pace, synchronized to a device called a Kilroy's stoking regulator that sounded off every ten minutes or so to alert coal stokers to reopen furnace doors and restart a fatiguing routine of slicing coal, raking out clinkers, and reloading the ovens. This regimen induced hard sweating and had an infernal quality to it. As the ship's historian Richard De Kerbrech explains, men labored here amid stifling coal dust, hot airless bunkers, and loud clamoring. They brought with them both a sweat rag to soak up perspiration and a large oatmeal can filled with water to stave off dehydration from the hours spent repetitively lifting heavy instruments like a devil's claw to rake out the fires or a forty-pound jumbo (an extra-long poker) to drag slag from the back of the furnace.[25]

This contrast between the furnace room and life upstairs could not have been more striking. A gentleman in the first-class cabins was expected to answer to the dinner bugle each night, freshly bathed, *en regal*, wearing a black suit and coat tie. Only a few flights below firemen, soaked in sweat, shirtless, or wearing uniforms of coarse-cloth dungarees, cheese cutter caps, and hard clogs, answered to a more aggressive gong, set to a quicker pace. For them, there were no mirrors for preening, no dressing hours, no extra accessories: life was intended for someone else's pleasure.

Modernity passes itself off as automated: but it was, and is, a sweating, exhausting, and ill-tempered thing.

The strict physical architecture of the *Titanic* divided the classes from one another, but that division was also reinforced by stark social inequalities. On a purely economic basis, the divide between the black gang and those in black tails was nothing short of staggering. Coal stokers on the *Titanic* would

Down in the boiler room. The USS *Massachusetts* fire room, late nineteenth century. *Edward Hart/Library of Congress*

have taken in wages amounting to about six or seven shillings a day (about forty-two dollars in today's terms without benefits, long-term health care, or security of employment). By contrast, to book the most exquisite first-class staterooms upstairs for a few days cost as much as fifty-six thousand dollars (in today's terms), or what would have amounted to about 3.6 years of labor, that is, assuming a coal stoker could find such work, stay healthy, and not take a day off. Heaped onto that material divide was a cultural condescension that relegated the coal stoker to a lower-class status in public life. No matter how hard a fireman worked, he was not expected to someday join the members of the elite in the first-class dining room. In fact, we can see that class divide played out in the recollections of one second-class passenger. He recalled stopping in Queenstown, just before the *Titanic* took its leave across the Atlantic. During the layover, he and a few people who were wandering around on deck caught sight of a coal stoker popping his head up through one of the ship's ventilation pipes. The man had climbed up a dummy funnel to get some fresh air, or maybe he did it just for fun. But that incident provoked social commentary. A few women declared it to be a "bad omen" and interpreted it (in their later recollections) as a sign of the "unknown dread of dangers to come."[26] Of course, we can't know why middleclass passengers would view a coal stoker's presence in this way, but we can certainly feel behind it the class prejudices, including a discomfort and disdain for the working people at the bottom of these commodity chains.

The life story of 03266 does not, however, point us merely to exploitation and exhaustion. If it did, we would not be living in a fossil economy. Shoveling, stoking, and breathing coal was, after all, not meant (at least from the perspective of the consumer) to do anyone injury. Injury was simply a byproduct. Coal was combusted because it produced enjoyment, wealth, and security for someone somewhere else.

Let us start on deck A—on the first-class promenade. Here coal purred rather than roared. My chunk of coal would have been felt in the first-class lounge or reading room located on this deck as a vibration, a soft buzz, or a "throbbing," as one passenger put. That is to say, one needed only travel a short distant from the boiler room to leave behind coal's oppressiveness and to start to see, feel, and hear its eroticism—to understand how modernity churns, generating strong attachments.

This eroticism that coal stoked was experienced, of course, along specific social axes of gender, class, and, ethnicity. Class was, in this respect, a key determinant on the *Titanic*. Edwardian culture reveled in the sharp social distinctions that today we brush under the rug. Not only did it have a first class, it had a second class and a third class, each of which had clearly demarcated spaces onboard the ship, policed by stewards, grilles, and social protocols. If nowadays we designate a few seats on the plane or railway to be first class and then euphemistically call the rest "coach," the inequality of fossil capital was more openly, and boastfully, displayed onboard the *Titanic*.

In this respect, the *Titanic's* first class represented fossil capitalism's structures of privilege. The ship's richest man, John Jacob Astor IV, symbolized the persistence of old wealth in the new economy. Like many others in the first class, his status was not earned or derived from his own labor. He was the great grandson of an immigrant fur trader who had amassed a fortune before the Civil War and who gone on to purchase much of Manhattan and its rents. His family's subsequent marriages into wealth had made it the richest in America and thus able to capitalize on, so to speak, the booming economy.[27]

More interesting, however, were fossil capital's nouveau riche who also sat in the first-class cabins. Their power was more directly derived from coal and oil and from the new structures of class exploitation that the mineral economy enabled. These people included Benjamin Guggenheim, who inherited his money from a portfolio of mining and smelting industries, William Ernest Carter, whose fortune traced back to the collieries and blast furnaces of Pennsylvania, George Widener, who built Philadelphia's electric streetcar system, George Wick of Ohio's Youngstown Sheet and Tube Company, and top railway executives like Charles Hays of Canada's Grand Trunk Railway and John Thayer of Pennsylvania Railroad. Among these, the Pennsylvania heiress Charlotte Drake Cardeza stands out as a fitting symbol.[28] Her wealth derived from her father's key role in modernizing Philadelphia's textile industry around steam power prior to the Civil War. That steam mill drove the labor of 210 textile workers, historically among the lowest paid workers in the fossil economy, and it ran 236 tireless looms and 10,000 spindles simultaneously to achieve economies of scale. Her story was thus the archetypal one of fossil capital: the capture of physical power commuted into social power.[29]

Our world pretends—at least in ideology—to be a classless, socially fluid space. Modernity revels in stories of rags to riches—of the Carnegies, the Clintons, and the Vanderbilts—and fossil capital justifies inequality through the possibility of movement. But it is worth reminding ourselves that the fossil economy both inherited the momentum of wealth and produced new stratifications that made immobility as modern as climbing up the social ladder. That contradiction stood in close quarters on the *Titanic*.

The role that coal played in reinforcing class stratification was twofold. First, coal had, as we have seen in the case of the ship's nouveau riche, empowered capital through a range of technologies and economies of scale that allowed it to refine its ability to accumulate the labor of others. That empowerment, symbolically encapsulated by the steam engine, reshaped the deep social structure of the world's economy, from financial portfolios to labor relations and across coal mines, farms, assembly lines, and boardrooms. If we take it to be true, as Karl Marx did, that wealth derives from extracting surplus from other people's labor, then what we see under fossil capitalism is that the world's first class, and even middle class, became decidedly more efficient at taking a little extra each day from the bodies and families of lower-class workers. So much seems obvious, if politically contested. But coal also played a second role on the *Titanic* by fueling, quite literally, the social performances of first-class life.

To start, material inequality was on stark display on the *Titanic*. Fossil capital is a highly stratified space, and the *Titanic* was one of its most spectacular sites of consumption. Much of that consumption was, of course, enjoyed by and reserved for the few. A coal stoker might arrive on ship carrying a few personal belongings (such as some coarse dungaree clothing, a belt, photographs, and a few odds and ends) and come prepared to sweat, but the *Titanic*'s first-class passengers came fully loaded down and ready to be served. The sheer material gulf between the classes was as arresting then as it is today, with the so-called 1% enjoying most of the perks of the fossil economy. Cardeza, for example, arrived with fourteen steamer trunks, four suitcases, and three packing crates, and her luggage included as many as eighty-four pairs of gloves, seventy dresses, jewelry worth more than a coal stoker's entire life earnings, and a train of boas, parasols, and ermine muffs. Similarly, her personal suite on the *Titanic* reflected the surplus upstairs relative to the austerity of the boiler room. Her suite, decorated in Tudor style, spanned three rooms and included maid's chambers, and it came with its own marble fireplace, extra sitting room, welcome bouquet of flowers, and an exclusive promenade of carved wood and greenery that ran for fifty feet of the ship.[30] Despite its proximity to the boiler room, the material difference was arresting.

These comforts were, of course, strictly privatized. For example, the entirety of deck A with its first-class smoking room, palm court veranda, library and reading room, and enclosed promenade running the length of the ship was exclusively reserved for the ship's richest passengers. So too were most of decks B, C, and D. Additional amenities for the first class, and in some cases the second class, came at a surcharge and included a fully equipped gymnasium, steam baths, heated pool, and squash courts located on the boat deck and the lower levels of the ship. All of this gave the *Titanic*'s privileged passengers a very wide latitude for leisure while in transit. Personal records indicate that they took advantage of that leisure by engaging in activities like relaxing to a cup of bullion and an ocean view wrapped in blankets on the promenade served by the ship's staff, taking a brisk walk on the promenade before the dressing hour with a companion in arm, and sitting back to a ten-course gourmet meal cooked in the ship's complex of kitchens. My piece of coal was, in other words, sublimated in these spaces of consumption even as its traces (in the form of congealed energy, ambient energy, mechanical energy, and embodied energy) could be seen and felt everywhere—through large arched windows, stained glass panels, and across trellises of ivy and oaken staircases.

The point is that for those at the end of this commodity chain, steam travel, or what we might think of as a collective automobility, did not entail work. Like fossil capitalism at large, it felt "splendid" and could even feel "gentle"—and it was an experience that nurtured choice, relaxation, status, and the perception of security.[31]

But my piece of coal would have also played a more direct role in fueling the rituals of first-class consumption onboard the ship. The deep energy of the *Titanic*—its dynamos, electric wiring, steam pipes, boiler rooms, and turbines—was there specifically to prop up the high-energy habitus of first-class living (and to a modified extent that of the second and third classes). Or to be more materialist about it: the ship's energy systems, its heat, light, refrigeration, ventilation, and mechanical energy, stood behind the comforts of its privileged consumers.

Embodying coal in the first-class suites and amenities was a business of pleasure rather than pain. Lighting is a case in point. The *Titanic*'s soft ambience, emanating from faux electric fireplaces and candle-shaped lamps, dripping from chandeliers, and glowing in electric torches held by cherubs, gave to the ship's alcoves, smoking rooms, and bedrooms a distinct aura of incandescence, a feeling of calm, of being first class. Similarly, the ship's microclimate controls, including its private bedroom heaters placed to ward off the chill of night, its hot and cold baths to soak the body in, its ventilation system that was designed to pump fresh air into cabins, and its heated swimming pool for tense muscles, were all designed to regulate, or modulate, the upper-class and bourgeois consumer's body to a temperature that induced ease.[32] Coal provided, that is, the first-class ambience to the ship's social performances.

The privileged body also felt coal's eroticism in various other ways that nowadays seem banal. For example, my chunk of coal fueled the *Titanic*'s complex of electric bakeries and stoves that allowed it to cook and cater six thousand gourmet meals a day for passengers and crew. No one upstairs was chopping wood or stoking furnaces. That energy derived from below. Roasted duckling, lamb in mint sauce, fillets of brill, corned ox tongue—these first-class delicacies, baked, boiled, and sautéed in the ship's kitchen complex—

Fossil capital's first-class ambience. *Public domain*

depended on combusting fuel below. That was also true of the *Titanic*'s immense refrigeration system. Here coal fueled electric larders and freezers for storing perishables, a cool drinking water supply, and an ice-making capacity that allowed for small pleasures like a chilled cocktail at the bar or a dessert of punch romaine. Such refrigeration ensured a first-class supply of fresh foods, vegetables, mutton, eggs, flowers, and milk, and it did so in quantities that were staggering—for example, seventy-five thousand pounds of beef, six thousand pounds of butter.[33]

Of course, the most tangible, or visible, way that coal produced security and comfort for passengers upstairs was through the mechanical energy, or labor, it performed. Electric cranes and winches to load heavy cargo like steamer chests on the boat deck, electric "lifts" to take the first- and second-class passengers up flights of decks, electric ventilating fans to move fresh air throughout the ship's chambers, electric gymnasium equipment, electric sorbet machines, and electric telephones. These converted coal into a type of disembodied labor, which produced the feeling of living an automated life wherein sweating was kept out of sight, a life that relegated the somatic costs involved in producing that energy to some hidden basement below.

To fully understand coal's eroticism, however, we need also to turn to its role in the rarefied—and more garish—rituals of first-class consumption. We can take as exemplary in this respect the *Titanic*'s spa, or what were called its Turkish steam baths, sitting just above the stokehold. Here the first-class body, stripped down to the skin, except for a swath of towels, was saturated in coal's combustion through an interplay of therapeutic heat, steam, and forced cool air. This routine began in an oriental-tiled room lit by electric Arab lamps where the customer disrobed, weighed in, and prepared for treatment. It then proceeded through a sequence of climate-controlled chambers that passed the body from a moderately dry heat room to a wet steam room and then to a hot room that burned at two hundred degrees Fahrenheit. Following that electric sweat, customers then recovered in a cooling room with a chaise lounge and a glass of refrigerated water before they went on to enjoy any number of other high-energy therapies. These included a horizontal water massage where the patron was subjected to pressurized injections from an overhead "blade douche," a 360° shower that pumped water down on the customer from all sides, and a bizarre "electric bath," in which the patron's body was placed in a capsule and bathed in electric lights intended to regenerate its energy supply.[34] Here was a therapeutic sweating of choice, a first-class sweat that contrasted sharply with the compulsory sweating happening only a few feet below.

It is hard to overemphasize the structuring role of coal in propping up first-class rituals on the *Titanic*. Oddly enough, even in those spaces where coal was absent, its combustion informed the habitus of life through negation. The pretext of coal, of leisured bodies in need of physical activity, was even present in those spaces where the first class chose to sweat without assistance, such as in the ship's gymnasium, squash courts, and long promenades.

The *Titanic*'s modern fitness facilities are a case in point. The ship's gymnasium was not intended to be used by those who labored manually. It was a first-class affair, a white-collar experience, meant to counteract the flaccidness and perceived neurasthenia caused by upper-class inactivity—an activity to keep privileged bodies, as it were, "in the pink of condition."[35] In this respect, the *Titanic* was no slouch. Its gymnasium contained state-of-the-art weight-lifting mechanisms and cardiovascular apparatuses, and its patrons also received the attentions of a personal trainer. Here one found upper-class men simulating the hard muscular labor of pulling oars on a rowing machine and lifting levers of heavy weights, and upper-class women riding stationary bikes, with the distance traveled metered by a gauge on the wall. Even so, coal also found a more direct entry into this sanctuary of the upper-class and bourgeois body. It famously offered consumers the chance to ride, for example, on an "electric camel" or "electric horse" in the gym, high-energy apparatuses that were meant to simulate the "exhilarating sensations" of riding outdoors and to circulate the blood of the neurasthenic body. First-class regimens of health had integrated, that is, electric instruments into their therapies of the flesh. The deep irony, in this regard, is that privileged bodies found themselves turning to energy inputs to counteract the dilatory effects of the high-energy world that had brought them here in the first place.[36]

In this stratified world of fossil capital, we might turn to the ship's first-class fireplaces, decorated with ornamental pieces of ceramic coal, radiating

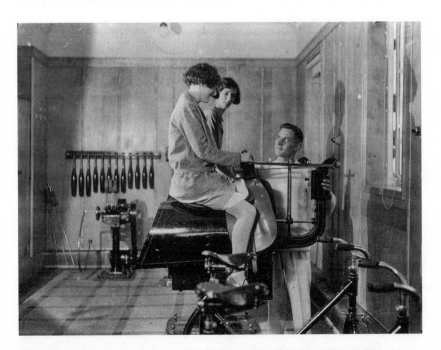

First-class bodies in the electric gymnasium, 1912. *Douglas Miller, Hulton Archive/Getty Images*

a clean flicker of heat, a gentle iridescence, to see how fossil capitalism works to sublimate its costs, to pass itself off as polished ease and plushness—a world away from the soil and precarity that fuels it forward not so far below. That repressed dialectic of security and risk still haunts us from the seafloor today.[37]

Of course, the first class was not alone on the *Titanic*. The ship also had a second and third class. These, however, require less treatment.

The second class was, in most respects, simply an ersatz version of the first. As in the world of fossil capital at large, second-class consumption manifested as little more than an imitative, and delimited, version of the conspicuous consumption upstairs. The ship's bourgeoisie might have been restricted from the most sumptuous experiences of the high-energy life, but it still felt my piece of coal as privilege and ease in its heated cabins, its pumped hot and cool water, its freshly ventilated air, its cooked food, its electric lighting, and the leisure it was afforded from the experience of traveling without labor. The real difference was that the ideals of bourgeois individualism and privacy (to which fossil capitalism's technologies had given such life) were not perfectly confident in these cabins.[38] Many second-class passengers had to book passage with strangers, share limited bathrooms with one another, and look on in envy or aspiration at those who enjoyed fossil capital's most prestigious comforts. Such differences meant that the bourgeois experience never quite matched the exquisite privatization of first-class consumption.[39] The social divide, however, grew to substantial proportions in the third class. While passengers in third-class steerage enjoyed basic amenities like fresh water, lighting, and heat, they were (like the ship's black gang) locked out from most of the *Titanic*'s leisure spaces, including its electric lifts, steam baths, finer dining facilities, smoke rooms, libraries, gymnasium, racquet courts, grand staircase, and long promenades. Although coal also pulsed here, it did so in more restricted ways, giving to the experience of fossil capital a less privatized and more utilitarian character. Passengers in this class slept in bunk beds with three or four other passengers, they ate in simplified dining facilities from a reduced menu, and they traveled not primarily with families for business or pleasure but as single men expecting to labor in the booming fossil economy across the Atlantic.[40] The *Titanic*'s third class occupied, that is, a social status under fossil capitalism that was only a step or two above that of coal stokers—and their experience of coal's combustion reflected that status.

Which is to say, class mattered in shaping the fossilized rituals onboard the *Titanic*.

The same was also true, if to a different extent, for ethnicity and gender. These also defined how passengers came to know and experience this institution of fossil capitalism.

For instance, the *Titanic* was not a deracinated space. Its physical architecture reinforced racialized divisions, and its launch was packaged within a racial imaginary. The ship's lead investors, for example, referred to the ship as an Anglo-American achievement—as the "pre-eminent example of the vitality and the progressive instincts of the Anglo-Saxon race." At least that is what its celebrants said over a dinner following the *Titanic*'s launch—to which we can only imagine a lot of white heads nodding.[41] Perhaps more importantly, the vast majority of the ship's first- and second-class cabins were overwhelmingly white and came from what today we call the global North, from places like Britain, Canada, and the United States. So much is not surprising, given its point of departure and arrival. That was also mostly true of the third-class cabins, although the latter was both demographically more heterogeneous and interpreted as being ethnically different, as being, at the time of the launch, a "happy" mix of the world's diversity or, when things went sour, a suddenly "dangerous" foreign element.[42] While the preponderance of steerage came from England and the United States, the composition of these cabins included a larger number of Irish, Bulgarians, Hungarians, Swedes, Finns, and Syrians as well. It represented, that is, the polyglot working class that was migrating during these years to do manual labor in the United States.[43] On the *Titanic*, the perks of combusting coal were thus parceled out along ethnic lines to a degree, even if that line of stratification was not as sharp or as stereotyped as that defined by class on the ship.

As with race and ethnicity, gender differences were also reinforced by coal's combustion. The *Titanic*—like Edwardian society at large—delineated consumer spaces along gender lines both formally and informally. For instance, its social architecture distinguished between the ship's more raucous smoking rooms, intended exclusively for men, and more feminized, if heterosocial, spaces, like the library and reading room, with its reflective incandescent lights and upholstered settees intended primarily for elite women. Although the birth of the New Woman had cracked a few of the older lines of division in some spaces of consumption, Victorian gender differences were still confident and hidebound. For instance, the ship's squash courts and gymnasium were still presumed to be primarily for men, even if certain hours were designated for women and children, and certain equipment, like a side saddle for the gym's electric horse, made the space more accommodating to the expectations of active women.

To be sure, however, coal upheld such gender performances primarily through indirect means rather than direct ones. The most concrete role coal played could be traced back simply to the larger phenomenon of mechaniza-tion and its impact on economic behavior. Over the course of the nineteenth century, coal and steam had made possible the expansion of the urban mid-dle class and, with it, the development of different assumptions about proper middle-class behavior. In particular, it had drastically reduced the need for manual labor on farms by improving work efficiencies per person. That freed a class of men and women from rural manual labor that once included harvesting and sowing crops, pumping well water to fill bathtubs and wash-tubs, scrubbing clothing by hand, and a range of other gendered duties char-acteristic of preindustrial life. By the time of the *Titanic*'s launch, such labor was not part of urban life, and the presumption among the middle classes had come to be that the idealized body, especially female ones, but even male ones, should be ornamental and free of sweat in public spaces. That was symbolized, on the *Titanic*, by first-class men arriving in black coat and tails, first-class ladies arriving in a different gown each night, bejeweled and wear-ing feathered hats, and second-class men and women doing their best to replicate that gender order and its impression of ease. Of course, gender per-formances looked very different in the ship's third-class cabins, where work-ing men and women could not expect to uphold such expectations of leisure and display. Consumption here had its own semiotics and character, even if it has received less attention.

What is clear is that combusting coal also stood behind the *Titanic*'s gender imaginary—with its ideals fulfilled in the first class, aspired to in the second class, and of only some relevance to the working class.

To conclude, my piece of coal—this artifact recorded in the books as 03266—had a rich social life. It had migrated through subaltern lives on the world's mineral frontier, it had moved through circuits of railways and docks, it had sat alongside laboring lives in the *Titanic*'s boiler rooms, and it was poised to pulse through the ship's sites of consumption. But, as we know, its social life never came to maturity. Three days after its launch, the *Titanic* struck ice and my piece of coal sunk, along with most of the ship's passengers, to the bottom of North Atlantic.

When Icebergs Strike: Fossil Capital on Ice

"Ice right ahead!"

That was the message sent out from the *Titanic*'s crow's nest at twenty minutes to midnight on April 15, 1912. The ship's watch had been on the lookout for growlers, bergs, and ice fields, common this time of year, but no one saw the one up ahead until it was too late. Seconds later, the momentum of this ship carried the *Titanic* into a force bigger than itself—into nature's unpredictability.[44]

Everyone on the ship felt the impact. Passengers in the forward cabins described it as an abrupt jolt, a jerk, or a thud, and then it was over. Just as suddenly, the world became quiet again—"everything seemed normal." Calm, clear, starlit. At least that is how some remembered it.[45]

The impact was more direct in the *Titanic*'s boiler rooms. Here the signs of impending system failure were almost immediately apparent. The iceberg had cut across three hundred feet of the ship's hull, creating a series of gashes, with the largest occurring in boiler rooms 5 and 6. Stokers in this section of the ship were consequently the first to understand, that is, to really understand, what was happening to them. George Cavell, a surviving trimmer, was working close to where the iceberg struck in boiler room 4. He recalled that the impact caused a large pile of coal to fall on top of him. He struggled to unbury himself and clamber out of the storage room to figure out what had happened. Even closer to the impact was George Beauchamp, a coal stoker, who was working in boiler room 6 when the ship was struck. Beauchamp couldn't see the breach, but he did see its effect. Almost immediately, water came gushing into that hold, and by instinct, he and his coworkers ran out of the room to avoid being locked in by the ship's automatic doors. But seconds later they were back at their station, working amid frigid waters that had risen up to their knees and waist as they attempted to draw out the furnaces. It took only ten minutes to reach a height of eight feet in the room, at which point that stokehold had to be abandoned. The same thing was happening in boiler room 5, only more slowly. The puncture in this hold was smaller and the sea came in as if by a fire hose. The stream here was gradual, but inexorable. Of course, that room too ultimately gave way to the Atlantic.[46]

The *Titanic*'s collapse—as it occurred in real time—was nonlinear, un-predictable, and, at some point, inevitable. Collapse always is. And yet, human resilience was strong in the face of disaster. Working men, for their part, struggled in the bottom of the ship with valor and discipline and with faith that a little human organization and agency might stave off system failure. For an hour and forty minutes following the impact, and in some cases for as long as two and a half hours, amid heavy steam, darkness, and rising waters, the black gang and the ship's engineers worked tirelessly together downstairs to save the ship. They bilged flooded chambers using hoses made of leather. They raked out furnace fires to prevent explosions in the remaining boilers. They kept the emergency generator running to maintain some semblance of safety and comfort. They scurried up and down ladders to deliver lamps to rooms where the lights had gone out. And they worked to save their coworkers by picking up injured men and opening doors for trapped men. It wasn't until 1:20 a.m. that they finally gave the ship up for gone and, at that point, with only a few exceptions, left their stations to climb upstairs to what they hoped would be safety.[47] Most of them, however, never made it.

There is a curious fact about the sinking of the *Titanic* that draws considerable attention. The lights stayed on throughout the disaster. Electric lighting radiated from the ship until only two minutes before it foundered. Some even recalled that a patina of haze went down with the ship as it sunk. The reason the lights stayed on for so long is that the emergency generator in the stern of the *Titanic* (and the boilers located there) had remained watertight up until the last. That meant that the dynamo kept spinning away, as if by its own ghostlike power, while the tragedy played out.[48] Modernity, as symbolized by this residual artifice of light, this vestige of the fossil order, proved persistent, that is, even in the face of failure. Of course, eventually the lights did give out when the ship cracked in half, exposing it to the sea, and opening up a vortex that sucked my piece of coal, and along with it most of the ship's passengers, down to the ocean floor.

Thereafter the world's energies became hostile, assembled against life itself, and survival became, once again, raw and primeval, coal-less, dependent only on starlight, luck, and the muscles of rowers. The fossil order had collapsed, and my piece of coal was no longer of use to anyone.

03266 sank to the bottom of the Atlantic with 1,517 passengers. And there is considerable debate about who went down with it and why. What was the social composition of catastrophe in this instance? Who survived and who didn't? Who got a life vest and who didn't when the lights went out? And, more importantly, why?

Might this be a metaphor of the coming crisis?

To be sure, people of all classes, genders, and conditions survived collapse. That much is known for certain. But, as we might expect, passengers did not all survive equally. Who you were and where you were when system failure came mattered. The composition of tragedy was shaped by gender, class, and position when the world finally collapsed.

In this respect, culture and character played a key role. In an age of gender inequality and enduring Victorian values, more women survived proportionally than did men. That is because the ship's protocols and its society's cultural mores offered life vests and seats on lifeboats to women and children first—at least in most cases. A certain gender ethics, that is, shaped how lives were valued and sacrificed when the ship went down. But it also gets messy very quickly. Not all women and children made it into the lifeboats. First- and second-class women survived at the highest rates. Third-class women did not. The main reason for that divide is that the ship's officers and stewards had to make fast decisions as the ship went down, and they did what they thought was best to maintain order. One important decision they made was to keep lower-class men who were gathered in steerage near the back of the ship from moving up to the boat deck. That was a decision based on a class unconscious that assumed working men and male ethnic others were more prone to disorder and thus potentially "dangerous." The consequence of that decision was that the women and children behind them could not make it forward, even as calls went out for them. While nearly all of the ship's first-class women survived and all but one first-class child survived, less than half of the third-class women and less than one-third of their children made it to safety as a consequence.[49]

Class and status also mattered, and to the extent that ethnicity correlated to class and condition, it too mattered. First-class men survived at the highest rates. Nearly one-third of the ship's wealthy men made it into lifeboats, even if most did not. In contrast, only one-tenth of the ship's second- and third-class men found a spot on those same boats. The reason as to why defies easy explanation in this case. Many of the ship's male survivors had simply found spots in the boats because the ship's officers assumed they were needed to row women and children to safety. Other men made it to safety because they were in the right place at the right time. They had shown up on the boat deck when lifeboats were still available, or they got lucky because no officer turned them back when they pushed through. In a couple of cases, men were simply strong enough to survive freezing water and climb into a lifeboat once they found one. One thing we know for sure is that most of the third-class men never got a spot in the lifeboats for the same reasons that third-class women and children didn't. The ship's authorities, in trying to maintain a semblance of order in the midst of collapse, had categorized them differently, moved them to the back of the ship, and held them back from the boats. In the face of collapse, that is, unconscious prejudices mattered.[50]

Perhaps most haunting in this regard is the memory we have of a group of lower-class men, of coal stokers and trimmers, coming up from the boiler rooms after the ship had flooded. Having worked to save the *Titanic* from collapse and having arrived on the boat deck very late to the game—with lifeboats and preservers in short supply, with the ship foundering, and with preferences being given to others—these working-class men were turned back, and, as if by fait accompli, ordered below deck where they, as they must have known, descended to meet their maker.[51]

Collapse is, of course, not a pretty thing, even if some of the *Titanic*'s surviving upper-class passengers were committed to remembering it otherwise. We hear, for instance, a belief in the civility of tragedy in one survivor's claim that "everyone met death with composure," that the passengers were, as she said, "a set of thoroughbreds." That was a story oft repeated.[52] But, of course, narrative is one thing and the mess of reality another. To understand what it meant to be there when fossil capitalism collapsed, we might tease out a different perspective from the memory of another survivor as she waited for the lifeboats. Standing with her husband on deck at a time when worry had not yet yielded to terror, she saw a coal stoker climb up from below. That changed her perspective.

> We saw a stoker come climbing from below. He stopped a few feet away from us. All of the fingers of one hand had been cut off. Blood was running from the stumps and blood spattered over his face and clothes. . . . I asked him if there was any danger. "Danger," he screamed at the top of his voice, "I should say so! It's hell down below, look at me. This boat will sink like a stone in ten minutes."

He staggered away and lay down fainting with his head on a coil of rope. At this moment I got my first grip of fear—awful sickening fear. That poor man with his bleeding hand and his speckled face brought up a picture of smashed engines and mangled human bodies.[53]

Black coal, smashed engines, mangled bodies. Collapse—even in cultural memory—is never a pretty thing.

In other words, a tragedy is never just a tragedy. It is many things at once. It is a commentary on a people's values. It is a testament to their luck and misfortune. It is a drama of their resilience. And it is, above all, a judgment on who we are—on what holds us together, what divides us, and what we are willing to do for each other when the world turns dark.

Today we again face crisis. Chunks of Greenland's ice shelf fall into the sea, broken off by climate change, drifting as icebergs toward dissolution in the lower latitudes. Tourists come to see them off the coast of Newfoundland not far from where the *Titanic* went down. But these are not the same

Fossil capital on the North Atlantic seafloor. View of the bow of the *Titanic* photographed by the ROV *Hercules,* 2004. *NOAA/Institute for Exploration/ University of Rhode Island*

icebergs. The disaster they portend is of a different character. Sea-lanes are opening up rather than shutting down. Sea temperatures are rising globally rather than, as they were the night of the collision, falling locally.

Once again, we have our eyes trained on the ice. We know the path ahead is obscured and uncertain. Growlers, bergs, and ice sheets begin to surround us. And there is still no reason to believe that someone else is looking out for us.

Three

Energy Slaves

The Technological Imaginary
of the Fossil Economy

Technology sits smugly at the center of the fossil economy. But our critical engagement with that technology is primitive, akin to carving meat with a chipped stone. We live in the fluorescence of the computer screen, climb into our cars, metros, and buses, yet little time is left in the day to do more than gripe or cheer over how these things, and the practices around them, define who we are as individuals and who we can become together.

This lack of critical engagement with our tools and apparatuses in today's consumer society creates certain blind spots that extend to our perception of the role that technology can play in climate politics. A trust in technological solutions (excepting a profound distrust in nuclear power) has pointed us to the engineer and the scientist as the answer to global warming. Salvation by innovation is, in fact, inscribed in two of the only three options we have available to us for averting a climax in fire and ice. A technological solution grounds, for instance, the option preferred by the world's privileged classes: that we replace fossil dependencies with either a safer version of nuclear power or a hyped-up version of solar power and wind power. Such a solution assumes that new innovations will permit life to go on as usual albeit with an infrastructural shift to a postcarbon economy. The second preferred option also leverages technology. It promises not a postoil or postcoal future but instead a doubling down on fossil fuels, a resolution to save the fossil economy by improving methods of carbon capture and carbon sequestration. This presumes only a slight reengineering of the status quo in that it calls for pumping carbon's consequences back underground. Of our options, it is only the third, and seemingly least viable, that avoids the technofundamentalism that drives these debates. This third option assumes that climate change is not a problem of engineering but rather a social and ethical problem rooted in inadequate values and institutions that prevent us from addressing the imbalances that go along with business as usual.[1] This option, which comes to us from the left and from the nondeveloped world, has had very little traction in the face of carbon's jubilee.

But if we are to act into today's impasse with any hope of success we need to get right with our technology, as communications critic Darin Barney puts it, to more clearly understand how it serves and injures us, and to do that we must, among other things, recover a genealogy of this fossilized self.[2] Such genealogical work, like all good history and all good therapy, means digging into and reprocessing the past so that we can see how it fuels the pains, thrills, compulsions, and satisfactions of the present.

To that end, let us start in an unfamiliar place.

Problematizing the Technological Fantasy: The World's First Mechanical Negro

The electric giant Westinghouse did a peculiar thing in 1930. It rolled out, to much fanfare and self-approval, the world's first mechanical slave, or what a southern newspaper called a "mechanical negro, at a National Electric and Light Association convention held in San Francisco."[3] This African American robot, which had been developed by the company's research laboratories, was one of the period's early humanoid automatons, a machine powered by electricity but made to look and act like a human. Baptized with the racist nickname Rastus by journalists, this black robot was meant to demonstrate just how far science and industry had advanced in perfecting a so-called mechanical man that might one day replace the human worker.

In one respect, Rastus Robot was an interesting showpiece. He stood up, he sat down, he swept the stage in front of him, and he uttered a few preprogrammed phrases to his audiences' delight. But from a technological perspective, that was about it. His work functions were strictly limited to a few such show tricks, making him of considerably less utility than the newest RCA radio, John Deer tractor, or GE refrigerator.

The value of Rastus was instead symbolic, or cultural. As one of the period's unique humanoid robots, he served, like the ethnic robot Katrina von Televox dressed in a Dutch maid's costume or the cigarette-smoking Elektro the Moto-Man, to give "a human face" to the terms *mechanical servant* and *slave* and to their successor the *electric slave*, these nineteenth- and early

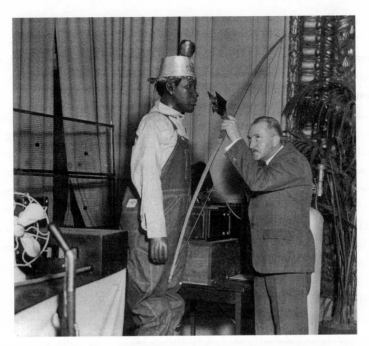

Rastus Robot, the "mechanical slave," built by Westinghouse in 1930. *Collection of Jim Linderman/Dull Tool Dim Bulb*

twentieth-century terms that denoted any labor-saving device that could convert energy into the functional equivalent of a laboring body.[4] News reports from the time suggest that Rastus did not disappoint in this respect. Journalists write that he simulated humanlike movements, that he spoke in "a rich Baritone voice" of current events, and that he performed in his signature act a rendition of William Tell by standing quietly on stage, apple perched on his head, waiting for an electric beam to knock it off. According to one report, at the climax to that act, at the dart of the arrow, Rastus reacted by sitting down in "dismay" and bursting out in a "human and startled cry" that made him seem just a little more like the rest of us.[5]

The cultural value of a robot like Rastus was that he personified mechanical energy. This mechanical body made to look like a human worker smoothed over the ontological chasm that had opened up between an earlier somatic world of working hooves, feet, and hands and the newly industrialized one that drew its labor power from these so-called mechanical slaves that provided labor without any of the humanity, the rambunctiousness, and the sentience of actual slaves or wage slaves. The term *robot*, Tobias Higbie reminds us, is revealing in this respect. It traces back to the concept of captive or "forced" labor (according to the *OED*), to the Czech word *robota*, and thus explicitly ties a mechanical body like Rastus to the social relationships between power and labor.[6]

Rastus was quite remarkable—in fact, startling—in this respect. His investor, Westinghouse, and his inventor, S. M. Kintner, had gone to great lengths to ensure that he not only looked persuasively human but that he represented a specific type of laboring body, in this case, an African American sharecropper. To achieve that effect, he was given a realistic physiognomy formed out of flesh-colored rubber rather than the more typical aluminum— with features of curly black hair, prominent lips, and rounded cheekbones— and he was dressed up to look like a field worker, outfitted in denim overalls and heavy work boots, signifying the standard dress of the cotton plantation. With a kerchief hung around his neck to wipe away imaginary sweat and to shield himself from an imagined hot southern sun and a tin pail placed upside-down on his head, in a peculiar indignity, this mechanical body not only looked like the descendent of an antebellum slave but also, as one contemporary put it, like a "dark-skinned minstrel."[7]

Black, male, and working class, Rastus was, in other words, different from other robots, and that difference gives us an opportunity to reflect on the founding features of the Western technological imaginary. He invites us, in a sense, to conduct archaeological work into the discourse of the mechanical slave or servant, these ur-tropes that elevate us to the presumed masters of technology (and the earth's energies) and that serve as the symbolic lynchpin in fossil capital's discourse of emancipation that has gotten us into such trouble.

The Genealogy of the Mechanical Servant and Electric Slave

The mechanical slave and its successor the electric servant first achieved widespread currency in the mid-nineteenth and early twentieth centuries, respectively, during the epochal rupture, or the "great divergence," that occurred in the world when coal and steam permitted industrializing societies to leapfrog over the resource constraints and muscular limits of the earlier somatic economy.[8] Similar fantasies had been around since the time of the Greek engineer Philo, who in the third century BC had dreamed up a robotic maid to serve his guests wine. But this discourse of mechanical servitude acquired its material grounding and began to circulate popularly much later in history. It acquired teeth, so to speak, during the rupture we know as industrialization when the now firmly entrenched fossil economy saw modernizing nations like the United States shift their energy base and labor burden (in the aggregate) onto subterranean minerals and the technologies that could transmute those minerals into the equivalent of a working body.

The social context of that paradigm shift was relatively straightforward. According to technology critic Despina Kakoudaki, the use of coal and steam engines to create labor power in English and American paper and textile mills in the mid-nineteenth century meant that workers and managers alike started to see and feel in intimate ways and in everyday practice an intensifying "exchange in the energies" between fossil-fueled engines and workers' bodies. That exchange, which began in the textile industries, quickly expanded into other industries and then intensified when electric technologies were applied to the assembly line in early twentieth-century factories.[9]

Together these developments, or what we casually term *industrialization*, created the historical context for the collapse of important ontological distinctions between living and mechanical bodies. For instance, the term *work*—this rich, dense, meaty, and sometimes suffering human signifier—came to be redefined in this period not as an exclusively human, or even animal, activity but as a generic thing—as the quantifiable product of the expenditure of energy (e.g., coal, food calories, petroleum) relative to the efficiency of a prime mover, living or mechanical, that could convert that energy into labor.[10] Work came to be seen as something abstract, the result of a force-producing movement, while the concept of energy was reduced to its current scientific definition—*the capacity to do work*. This reconceptualization of the material world and the human experience through a thermodynamic lens redrew the lines between living bodies and inanimate forces as well as between food and rocks, and it put steam engines and servants, minerals and slaves, iron horses and panting horses in uncomfortably close quarters.[11]

To get clarity on these matters, we might step back in time to 1827 and to an unsuspecting event, a special hearing held by Britain's Privy Council focused on the manumission of slaves in what is now British Guiana. Here we can catch a glimpse of the emerging equivalence between carbon technologies and human labor as it first took shape. In that hearing, a planter from Berbice who owned over fourteen hundred slaves, Hugh Hyndman, was questioned by the council regarding the equivalence between human labor and steam power, that is, between literal bodies and mechanical ones. The question was direct: "Is it a fact that steam is now employed in Berbice in substitution of labour formerly executed by slaves?"[12]

Hyndman's answer to that question was not surprising, and it reflected the degree to which mechanization had advanced in agriculture. Steam power had not, he said, made many inroads into the sugar plantations. West Indian slaves still planted the crops, picked the sugar, carried the cane, boiled the cane, and delivered it to port. In fact, as far as he knew, coal and steam were only being used in grist mills for breaking and grinding the cane, and this phase of the production process had previously not been done by humans but rather by "wind and cattle." Steam power might have "economized" slave labor in one part of the process, he said, but nowhere had it replaced human bodies.[13] Thus despite the proliferation of steam power in this period, there was as yet no functional equivalency on the sugar plantation between slaves and steam engines.

What is interesting in this early case, however, is that this discussion of steam power was introduced into a conversation over the "abstract value" of biological (in this case, slave) labor. England was at the time trying to determine what the proper monetary compensation should be for a slave owner under the proposed new emancipation laws. The steam engine appeared here as an example of how it was possible in theory for any given laboring body, following manumission, to be replaced by some other equivalent type of purchased labor power, whether another conscripted human body (but not that particular one) or some form of mechanical energy. The debate circled around, that is, the concept of abstract, standardized, and replaceable labor.

A defender of the old regime, this slaveholder insisted that enslaved labor was too variable to be flattened into one generic measure of productivity or monetary value. Hyndman argued in a perverse logic of self-interest that slaves were individuals and thus not alike in their capacity to be productive. Some, he said, were "more valuable for physical strength" in the fields, some brought unique technical skills, and others, he said, were valuable to the slaveholder because of "their moral character . . . and influence on the rest." By his calculation, only about one-third of the slaves on a plantation were actually what he called "effective and able" laborers, that is, adults able to perform physically demanding work day in and day out. The rest were, he thought, too young, too old, too resistant, or too disabled to measure up to that standard. Moreover, some slaves, like boilers and machine tenders, were not easily replaceable; unlike field hands, whose value lay in their physical might and endurance, these other slaves had technical skills that had a multiplying effect. These considerations made it difficult to put a definitive monetary value on slaves in the abstract. "I do not think," he testified, "that I could possibly put a just value upon the loss the master would sustain" by losing any particular slave.[14] Hyndman was, in a sense, simply stating the facts of the time: there was not, except in wildly gross terms, any accurate way to gauge in standardized units what was coming to be called manpower.

What is revealing here, however, is that coal and steam popped up in this discussion of the abstract value of human labor. Although the trope of the mechanical slave was not in play in this case, the detour that the conversation took was an important step in its formulation. The possibility that labor might not only be objectified (as it had been throughout history) but also abstracted and reducible to a standard measure, whether derived from a machine or a living body, and thus separated from the sweat and sentience of people was evident, in other words.

Hyndman's logic was not unique. It was part of the larger revolution in our understanding of human labor provoked by industrialization. By the 1820s that revolution had, in fact, also produced the curious concept of manpower, a generic analog of steam-power and horsepower. To quote one contemporary cited in the OED's entry for *manpower*, "The human body, as a whole is a self-moving, self-supplying, steam-engine, not of a horse-power, but of a man or woman-power." Although no one had yet surmised a scientific gauge for tabulating what exactly constituted a manpower unit, that development—a key step in the formulation of later debates over these energy slaves—was not far off in this effort by a liberalizing state to equate human workers directly to carbon technologies.

For a second example, we might push forward a half century to an urban industrial geography. By this point and in this place, the trope of the mechanical slave had taken firm root. The commentary of French journalist Eugène-Melchoir de Vogüé on the 1889 Paris Exposition shows just how much "the form of labor," as he put it, had changed, along with the perception of that labor.[15]

What most impressed de Vogüé at the Paris Exposition was the vast mechanical "organism" he saw operating throughout its exhibits. Gazing down from the balcony of the Palace of Machines, with his elbows on the balustrade, he imagined that he saw the "universal law of labor" revealing itself everywhere in the fair's complex of steam- and electric-powered machines. "This crowd of automatons," he said—these analogs of biological labor—seemed to him to be working tirelessly throughout the fair with none of the "fatigue . . . noticed in the arms of the [human] laborer." Their "mechanical arms," he wrote, "work the metals, weave the cloths, prepare the food, light the lamps; they sew, print, engrave, sculpture; they adapt themselves to all demands, the heaviest and the most delicate." Moreover, these carbon-powered laborers were the perfect servants. Their temperament was obedient and docile: they submitted to being "shackled," they "obeyed," and they went, he said, to "sleep" when ordered to do so. To de Vogüé, the lesson of the exposition seemed clear: "Man," he said, "now hides himself behind a great mechanical slave."[16]

De Vogüé's report on the Paris Exposition exemplifies how the language of organicism and mastery, which had historically been used to describe living labor, was coming to be applied to machinery in a way that evaporated former ontological distinctions of importance. What was not exactly true, however, was the conclusion that he drew from these developments: that mechanical slaves had *replaced* actual slaves and other manual laborers in the modern world—or, more precisely, that it was simply a matter of time, a technicality, that they had not yet done so. For de Vogüé and the industrial bourgeoisie at large this trope of the mechanical slave, that is, performed ideological work, serving as the centerpiece in a narrative of technological mastery and emancipation that came straight out of the Enlightenment and served class interests.

De Vogüé's observations of an African traveler standing next to a steam engine make clear how the trope of technological servitude operated in that narrative. Having chanced upon this Sudanese traveler in one of the fair's engine exhibits, de Vogüé imagined him to have been emancipated by carbon technologies: "Occasionally," he said, "a Negro from Soudan is seen in this gallery examining a steam engine. He would get down upon his knees before it, if he knew how much he ought to bless this *mechanical slave* which has been substituted in his place." The language was insensitive if strong. This "black flesh," which had been replaced by coal and steam, he wrote, augured a fully liberated world wherein the mechanical slave—this "deliverer of the human race from the most irksome forms of toil"—would replace manual labor at large. Mastery of nature—in this case, unlocking the labor potential in prehistoric carbon—was, in other words, the material basis of modern freedom and equality and the imagined mechanism for ending actual slavery.[17]

In such a context, the mechanical slave was a glib, even frightfully crude, metaphor that misrepresented the role that coal, steam, and electricity had played in the world. But it also had enough truth in it—after all, the form, intensity, and scope of mechanical labor had revolutionized society, and such mechanical labor frequently displaced human labor—to retain a hold on the popular imagination and to serve as the linguistic lynchpin in this "religion of technology" that was becoming so engrained in bourgeois culture, state propaganda, and industry boosterism.[18]

The Social Unconscious of Technological Servitude

This discourse of technological servitude also reproduced a certain social bias, a class, racial, and gender positioning, that is instructive to reconstruct. On the one hand, contemporary references to mechanical servants and slaves in the nineteenth and twentieth centuries were typically cast in generic, or deracinated, terms to indicate the mechanical liberation of the nation, or the species, writ large. For example, state propaganda coming out of agencies like the Rural Electrification Administration referred simply to having access to "the cheapest servant" one could buy and learning to relax since these "kilowatt hours don't get tired."[19] Likewise, Gainaday Company, a consumer goods company that sold electric clothing wringers, marketed the notion that a person might be emancipated from work—that she might "gain a day" free from muscular labor—by bringing into the house what it termed a "dependable servant" and "inexpensive worker."[20] And the once familiar trademark figure Reddy Kilowatt, a blue-collar cartoon character with arms, legs, and a head attached to a bolt of lightning, introduced by an Alabama electric company in 1926, was pitched simply as the nation's faithful "electric servant . . . who worked for pennies" so that the rest of us might be free.[21] Here the language of servitude was generic and evacuated of its history.

But the rhetoric of technological servitude and its prevailing discourse of mastery reproduced a decidedly classed, racialized, gendered, and sometimes regional positioning, with the repressed referent of the subaltern body and especially that of captive black labor not far below the surface. This is not surprising, because the privileges of the high-energy society were, and are, socially stratified along such lines. Racial inequality, for example, provides a barometer. In 1940, the vast majority of white Americans had electric lighting in the home (82.9%) whereas the majority of nonwhites were still fueling lamps by hand with kerosene or gasoline (55.2%). Additionally, nearly half of all white Americans had electric refrigerators in contrast to the majority of the nation's minority population, which had no refrigeration or simply used ice blocks. And if most white Americans had gas or electric stoves by that point (58.7%), nearly three-quarters of the minority population was still stoking stoves by hand using coal or wood (70.8%).[22] It is especially telling that even today African Americans have only 55% of the median equity in the automobile culture of their white counterparts, a much lower investment in fossil capital's marquee energy technology and its expansively subsidized infrastructure.[23] Of course, disparities are even more extreme outside of the developed world, where 1.1 billion people living in sub-Saharan Africa, parts of Asia, and rural areas lack even rudimentary electricity.[24]

The uncomfortable racial, class, and gender unconscious of this discourse can be seen close to the surface in a 1924 article entitled "Idle Slaves of the South." That article, written by a southern modernizer named Marion Jackson in the peak years of electrification in the United States, concerned the enormous hydroelectric potential of this underdeveloped region. Written at a time when a racialized system of sharecropping was still in full swing in the region's cotton fields, Jackson's article looked out onto the untapped energy of the region's rivers and streams and saw not simply untamed mechanical power nor generic mechanical slaves but quite specifically underemployed black bodies—"idle . . . Negroes," as he put it. Here in the South's falling waters, he said, were "millions upon millions of slaves . . . offering to serve, waiting only the command of man," a veritable army of black bodies "equal in labor energy to more than 75,000,000 men [or] over five times the Negro population of the United States."[25] Given its historical context, the analogy struck very close to home, evoking the de facto structural racism of the region.

The historical thickness—and the racial positioning of the audience as white—gave this analogy of mechanical servitude its rhetorical and affective force. While there was nothing to prevent such metaphorical slaves from working on behalf of the African American community or any other minority community, the terminology of liberation, mastery, and service typically highlighted the uplift of white bodies and directed its appeal to white audiences. That rhetorical positioning, in other words, replicated the de facto material stratification of property and opportunity in the high-energy society.

Equally important was the gendered logic that gave this discourse an emotional tug. Jackson spoke, for instance, of the indignity of America's rural women grinding out their days in strenuous physical tasks like chopping firewood and carrying water for purposes of cooking, heating, bathing, and cleaning. This type of backbreaking labor—symbolized by what he called "the gaunt farm woman lifting daily her ton and a half of dead weight"—did not align neatly with middle-class visions of white women's role in society, and it fueled the technologist's dream of liberating the nation through energy development.[26]

In this instance, the concern was specifically with white women and their families living in a prostrated South that Jackson said had never recovered from the fall of slavery and that he thought might be resurrected on a new foundation of mechanical slaves. "Americans of the purest strain," he wrote, these beleaguered white bodies living amid "drudgery" and "darkness," might be returned to their rightful rank once these millions of metaphorical "Negroes" were put to the work that actual slaves once did.[27] Of course, that fantasy was a peculiarly regional one, tied to a specific time and place, but it speaks to the fact that this broader discourse of technological servitude was never socially or racially neutral.

A second article from *Popular Mechanics* entitled "Thirteen Slaves for a Nickel" helps to further delineate the racial, class, and gender dimensions of this metaphor. Written in 1939, in the midst of the long fetch of the consumers' republic, this article tied the joys of a midcentury high-energy lifestyle to an earlier history of racial and class privilege. Here the pleasures of the white middle class having work done for it by these electric washing machines, furnaces, dryers, stoves, and other mechanical devices—these modern "slaves"—was tied up with a racialized nostalgia for the plantation and for having actual servants and gendered black bodies at one's beck and call.[28]

"You look back," it started, "with a twinge of envy to Mount Vernon, and lament that the ease and splendor of that graceful period of American life are vanished forever." The article then turned the tables: "Don't feel sorry for yourself. You are a bigger slave owner than George Washington."[29]

The premise in this case was that Washington, living at the height of eighteenth-century fashion, had only 66 actual slaves (in reality, he had 123 plus additional dower slaves and rented slaves) working his Mount Vernon estate, whereas the modern middle-class consumer had several times that number of mechanical bodies at service in the home, as many as 400 potential "slaves" which when activated could work at a "muscular rate" that far outperformed that of real bodies. These included "man servants," who kept the water hot and tended the furnace at night, "maid servants," who mixed, ground, and baked one's food with only the touch of a button, and "kilowatt coolies," who worked for the "lowest wage" at whatever other mechanical tasks could be shifted onto electricity.

This framing of access to technological servants by way of class-inflected, gendered, and racially somatic terms was drawn out in several other ways. For instance, the article detailed the various muscular tasks, like churning butter, woodworking, and distilling wine, that slaves did on Washington's plantation and guessed at the number of man- and woman-hours devoted to each of those tasks. It then estimated that the work being done by mechanical energy in the average suburban home—in the form of tasks like pumping water, heating rooms, washing clothes, scything grass, blending fruit, and lighting bedrooms—added up to as much as "156 servant-hours a day." Moreover, it suggested that whereas the master of Mount Vernon had to bear the additional burden of housing, clothing, and feeding his "scores of negroes," modern consumers of electricity could own these "husky, versatile, and willing slaves" for less than a cent an hour without worrying that they might "complain," "slam the door," or "gossip" behind one's back.[30]

What is revealing in this rhetoric of mechanical servitude is that the imagery of the white middle-class body, and in particular the white female body, being serviced by servants, coolies, and slaves was so explicit.[31] This rhetoric evoked an emotionally charged history of black men's bodies sweating in southern cotton fields, of Asian men's bodies toiling in Western laundries and colonial plantations, and of black and immigrant women serving their mistresses and masters in white middle-class and upper-class homes. It thus made transparent the otherwise tacit racial, class, and gendered substructure that gave to the trope of technological servitude much of its rhetorical force and that replicated in ideology the realities grounding the social stratification of the growing consumer's republic. In such a context, the advent of a "realistic mechanical negro" like Rastus seems culturally logical, if abhorrent, as it speaks to the historic inequalities and blind spots that evolved in the nation's technological imaginary.

Quantifying Mechanical Servitude

How this discourse of mechanical servitude acquired a quantitative grounding is equally important, since its quantification has shaped how we understand the scale of the impact of carbon technologies on modern societies and the human experience. In order to gauge that scale, someone had to deduce what the abstract work output of a human body might be in kilowatt hours or develop a common measure that would allow the output of inanimate motors to be translated into that of the so-called human motor. That is, what constituted a generic "manpower unit" had to be calculated, along with the energy calories required to produce that unit of manpower.

Quantifying biological and mechanical labor, in this modern way, was undertaken as early the 1780s, when James Watt attempted to determine the horsepower capacity of his steam engine by running a set of calculations based on the ability of draft horses to perform comparable mill work. But it became a much larger business in the mid-nineteenth century when physicists, physiologists, and engineers in Europe and North America, who were grappling with the changes wrought by mechanization, reconceptualized the human body as an engine within a new physical paradigm that located energy (i.e., calories) and its conversion into work as the cornerstone of modern scientific thought. An entire industry of body studies developed in these years in the work of European engineers and physiologists like Etienne-Jules Marey and Hermann von Helmholtz that relied on time-motion studies, studies of respiratory rates during physical exertion, and studies of metabolism and fatigue to offer fairly precise calculations of the rates of energy efficiency, mechanical capacity, and physiological limits to organic bodies. This work in rethinking the body as a "human motor" capable of generating a standardized work output reached across the Atlantic and became popularized in the writings of industrial engineers like Frederick Winslow Taylor.[32]

We might take as representative the proceedings of the Franklin Institute, one of the nation's oldest scientific organizations in Philadelphia. Comprised of engineers and academics, this group gathered in 1873 to take stock of engineering terms concerning work and power and how they were being used and abused by various parties. In a treatise entitled *Establishing Precision to the Meaning of Dynamical Terms*, the members of the institute proposed to nail down the definition of *manpower* to fifty-foot-pounds per second of power, or the force needed to displace fifty pounds by one foot in one second—a figure they took directly from Arthur Jules Morin's earlier dynamometer experiments estimating the work output of various bodies in motion. They further proposed to lower Watt's estimate of a horsepower unit from 550 foot-pounds per second to 500 on the grounds that it overestimated actual horsepower and that the lower figure made for a tidier one to ten ratio of manpower to horsepower. The only additional rationale they offered for that latter ratio was that a man weighs about one-tenth that of a horse and that he eats approximately one-tenth the calories.[33]

This generic concept of manpower reproduced the assumption that work meant merely physical force rather than skill, precision, or intelligence and thus proposed as its gauge the adult male body at brute tasks rather than some other measure. Such a redefinition of work, of course, elided the complex social realities of manual labor. Subsequently, however, one manpower unit, translated into wattage, came to be calculated simply at seventy-five watts of power, a consensus that, while always questioned, has remained intact through today.

With such terminology in hand, geologists, economists, and engineers tossed around some ballpark figures of the nation's total mechanical servitude in the early twentieth century in an effort to quantify the material progress that had been made in adopting carbon technologies. Those estimates derived chiefly from national inventories of mechanical horsepower capacity taken by the US Geological Survey, which tallied the nation's prime mover capacity in automobiles, steam-powered railways, electric stations and trolleys, manufacturing engines, carbon-powered ships, and farming and irrigation technologies. The estimate put forward by the director of that survey, George Otis Smith, was that Americans possessed, even as early as 1921, approximately a billion manpower units in mechanical labor, or, what he described as ten "energy servants" apiece "to do our bidding."[34] Other contemporaries, who tweaked his assumptions a bit, arrived at figures as high as three times that number—a total of "three billion hard-working slaves," or thirty mechanical slaves for every "man, woman, and child" in the United States.[35]

Of course, none of this was a perfect science. The disparities make that clear. The numbers changed depending on assumptions. Should mechanical equipment not in use, or periodically used, still be counted as a full horsepower? Was it sensible to assume a manpower unit could be sustained in the same way as that of a fossil-fueled tractor? How was energy used for the purposes of generating power to be disaggregated from energy used for non-work purposes like lighting, heat, or cooling? Still, this earlier work set the stage for later assessments of the role that fossil fuels and technology played in modern life.

Out of these efforts to quantify mechanical labor and to assign to it a historic analog came the term *energy slave*, a trope, or meme, that now circulates in environmental debates, albeit in a way different than it did in its original instantiation. First used in 1940 by the popular intellectual Buckminster Fuller in a *Fortune* magazine graphic, this terminology of the energy slave shifted the thrust of the metaphor from an emphasis on machines to an emphasis on the energy consumption that fueled those machines. If earlier geologists, engineers, and economists had based their estimates on actual inventories of existing mechanical capacity, Fuller had simply translated all of our energy consumption, despite its end use, into theoretical manpower, or energy slaves. To derive his figures, he took estimates of national and global energy consumption at large, then calculated how much work potential was in that energy, and finally divided that number by the figure used to define a unit of manpower, seventy-five kilowatts. Bracketing the fact that as much as half of the world's use of energy went into heating rather than mechanical work, Fuller calculated that the world's two billion people now shared the globe with another "36,850,000,000 inanimate *energy slaves*" and that as many as 54% of those slaves, or "an army of 20,000,000,000," were housed in the United States alone. This, Fuller said, represented "man's answer to slavery," and it was driving the world's economy and populations radically upward.[36]

A final step in quantifying this concept of technological servitude was the stamp of approval given to it by the academy in the midcentury work of the anthropologist Leslie White. White's pathbreaking 1943 essay "Energy and the Evolution of Culture" provided this discourse with a recognized narrative format that while always questioned, qualified, and critiqued has persisted throughout environmental writing in the subfields of energy history and ecological anthropology.

White argued in this and a series of related pieces that our species' historical progress could be measured, at bottom, as the product of the quantity of energy consumed by a society multiplied by that society's technological capacity to transform that energy into work. He argued, in a now familiar story, that human history proceeded from low-energy strategies like hunting and gathering, where individuals had mostly "the energy of [their] own body under . . . control," through the stages of pastoralism and agriculture, which increased our control over solar energy through biological converters like domesticated animals and plants, to today's high-energy societies, where technologies for turning coal, oil, natural gas, and atomic power into work had revolutionized productivity, accelerated material wealth, and led to more complex forms of cultural development. That argument redefined the "cultures of mankind as a form or organization of energy" (relative to technology), and it reinterpreted all "behavior, whether of man, mule, plant, comet or molecule . . . as a manifestation of energy." "Energy is energy," White declared, "and from the point of view of technology it makes no difference whether the energy with which a bushel of wheat is ground comes from a free man, a slave, an ox, the flowing stream or a pile of coal." Here the human worker was shrunken down to that of a prime mover, to "a groaning and sweating slave" capable of generating "75 watts of power," and human culture was reimagined as nothing more than the quantity of material goods that resulted from converting energy into economic productivity.[37]

White's conclusion was as facile then as it is today. He concluded that "cultural development," or "C" could be measured by simply taking a society's energy consumption ("E") and multiplying that number by its technological capacity to convert that energy into work ("T").

To wit:

$$C = E \times T.^{38}$$

Or to paraphrase: fossil fuels multiplied by technology equals modern culture.

Such a bold, and simplistic, conclusion was, of course, not tenable, but it demonstrates how the concept of energy as slave to humanity came to fit within a formulaic structure that embedded it in a master narrative of human evolution that assigned to technology and humanity's mastery of nature's energies an uninterrogated role. By such a gauge, the mechanical slave, the body of Rastus, or the "T" in the equation was assumed to outweigh all other nonmaterial factors in the progress of our species. That assumption and modified versions of it have proven hard to dislodge.

The Energy Slave in Climate Discourse

To be sure, Rastus, along with this discourse of technological servitude, has not gone away. It sits comfortably in the West's technological imaginary. For instance, even environmental activists have resuscitated Rastus's body—now termed an *energy slave, virtual slave,* or *ghost slave*—to critique modernity and its addiction to fossil fuels. But they, myself included, have done so in a way that replicates some of the term's previous shortcomings and introduces new problems into the present politics of climate change.

The term *energy slave* popped up in its new environmental context sometime in the early twenty-first century, and its usage has since increased. The term has appeared in popular media outlets like MSNBC, PBS, Wikipedia, and Grist.org, and it has shown up in academic writings on the environment including such books as *Ecological Economics* (2005), *The Second Law of Economics* (2011), *Cultures of Energy* (2013), *Oil Culture* (2014), *Routes of Power* (2014), and in my own *Carbon Nation* (2014). The most influential use of the term derives from J. R. McNeill's landmark 2001 text, *Something New Under the Sun*, a widely-assigned environmental history of the twentieth-century world. In that text, McNeill argues that fossil fuels and their conversion into power, these *energy slaves*, so to speak, have played a singular role in propping up the material infrastructure of modern life and in defining the problematics of our present. As McNeill puts it, the average global citizen has come to depend upon twenty "energy slaves," or about "20 human equivalents working 24 hours a day, 365 days a year," to support the world economy and the material standards of living it makes possible. He tempers that claim by explaining that while the quantity of power unleashed by modernization was catalytic—"something new under the sun"—it was also cut through by differential access to that power that reinforced old forms of inequality and produced new forms. The average American might command, he notes, "upwards of 75 energy slaves," but the average Bangladeshi still depends on "less than one."[39] This rhetorical intervention brought a measure of self-consciousness to bear on our carbon dependencies, and subsequent environmental writers have found that rhetoric to be compelling.

Several premises are explicit or implicit in this logic, and each carries with it certain problems. To quote technology and science critic Sara Pritchard, the discourse is not entirely "innocent" on either ethical, historical, or political grounds.[40]

The first premise is a familiar one. It is that fossil fuels are equivalent to slaves, that when converted into work, they perform the same type of physical labor that slaves and other exploited bodies once did. This logic, seen above in McNeill's writing and that of sociologist William Catton's *Overshoot* (1980), has been rapidly duplicated across environmental discourse. We see it foregrounded, for instance, in the title to the award-winning journalist Andrew Nikiforuk's book *The Energy of Slaves: Oil and the New Servitude* (2011), which builds on the premise that each barrel of petroleum is "the equivalent of 3.8 years of human labor" and that modern consumers in oil-rich North America are thus dependent on "89 virtual slaves," or for "a family of five . . . nearly 500 slaves," to support their standard of living.[41] That claim appears casually across environmental writing, and it is not inert. It can be taken to mean, as one critic has pointed out, that petroleum and coal serve not only as the functional equivalent of laboring bodies but that they have caused slavery to be "a less desirable solution to global labor demands."[42] In other words, this premise can imply causality, whether intentionally or unintentionally, by suggesting that energy slaves had something to do with the displacement of actual slavery and even sweated labor itself. But as the historian Klas Rönnbäck has shown, the historic relationship between fossil fuels and slavery does not lend itself to so easy a conclusion.[43]

The second premise is that consumers living in North America and Europe represent a new type of master class that derives its social power from these "energy slaves." This premise is evident, for instance, in recent remarks made by anthropologists Thomas Love and Michael Degani in *Cultures of Energy* where they explain that even at the high price of one hundred dollars a barrel for oil (which translates into 450 weeks of human work), Americans have come to rely on some "very cheap slave[s] indeed." This easy access to mechanical energy, they say, inclines those of us who hold that power to adopt the same "demigod sense of self" and "unwarranted belief in our prowess" as that of the preindustrial slaveholder.[44] The strongest statement of this position comes again from Nikiforuk, who contends that the world is run by a "petroleum slaveocracy." He writes that Westerners are a privileged class that "hoards" its energy slaves and the fruits of their labor by gleefully burning up for its comfort as much as 23.6 barrels of oil a person annually without regard to the social and environmental costs of that practice, while the rising class of "new and leaner slave masters" in Asia accomplish their own mastery with only 2.4 barrels of oil a year, or, in his words, with about "9 coolies at their beck and call."[45]

The ethical implications of this comparison came to light in an article published in the journal *Climactic Change* in 2011. In that article entitled "Past Connections and Present Similarities in Slave Ownership and Fossil Fuel Usage," the historian Jean-François Mouhot argues that it is politically useful (if not also a bit risky) to compare past slaveholding to the burning of fossil fuels in historical and moral terms. We behave "like slaveholders" in the West, he remarks. "We are as dependent on fossil fuels as slave societies were dependent on bonded labour, . . . [and] in differing ways, suffering resulting (directly) from slavery and (indirectly) from the excessive burning of fossil fuels are now morally comparable, even though they operate in a different way." Mouhot is, of course, trying to shake things up a bit, and his rationale for doing so is that the cost of burning fossil fuels in the West is having and will continue to have such an adverse impact on the global poor—as a result of rising sea levels on tropical islands, extreme droughts in already arid regions, and unpredictable weather patterns in vulnerable coastal communities—that burning them with a conscious understanding of their costs has become morally, and historically, comparable to the previous exploitation of slave bodies.[46] Although most writers do not reach that same conclusion, their use of the analogy depends on a similar guilt by association.

The third, and final, premise is that premodern slavery was a type of energy system. This premise, which grounds the field of energy history, can be seen, for example, in Jean-Claude Debeir and his collaborators' increasingly influential *In the Servitude of Power* (1991). The authors of this text explain that the slave's ability, and more generally the human body's ability, to convert energy calories into labor makes those bodies functionally equivalent to watermills, windmills, and steam engines for the purposes of productivity. That logic places the practice of slavery under the category of energy management, or "the social control of energy," and it makes slavery historically equivalent to later practices of hydroelectric development or power grid management. Thinking within this paradigm can yield an uncomfortable panache. For example, it allows for the conclusion that previous cultures depended on "the slave *engine*," or that they operated on "the two-cycle *engine*—the energy of solar-fed crops and the energy of slaves," and it can

imply that slaveholders saw the world through the same instrumental ratio-
nality as the modern engineer.[47] Nikiforuk claims, for instance, that slaveholders "calculated the calories burned by slaves" with the "economic cool-
ness" of today's engineer, that they worried about energy losses (e.g., runaway
slaves and stolen food) in the same ways that the accountant does, and that
they were, like the efficiency specialist, hell-bent on recapturing lost energy
by hunting down fugitive slaves and punishing idle ones. Historical difference
gets flattened in this rhetoric when we hear, for instance, that the Middle
Passage was "a leaking pipeline," a wasteful energy practice that saw as much
as one-fourth of its potential energy lost to poor transportation practices.[48]

This language is meant to provoke, as historian Christopher Jones ob-
serves, and it has a catalytic verve to it that the drone of climate change
statistics does not.[49] But it brings with it some undesirable luggage that is
both old and new, and it is to those concerns that we finally turn.

Technology, Nature, and the Ethics of Energy Consumption

The ghost of Rastus haunts the nation's somatic past, present, and immediate future. Although long since decommissioned, he stands in for the
modern problematics of energy, technology, and labor. To conclude, we can
draw three lessons from the body of Rastus as we assess how this discourse
bears on today's public debates over fossil fuels and climate change.

First, the body of Rastus returns us to the increasing exploitation of
human labor under fossil capitalism.

The mechanical slave, this metaphor of disembodied labor, and the later
deracinated vocabulary of the energy slave have a tendency to showcase the
newest technology while underplaying the lives of humans and our historical continuity with the somatic economy. What is often missing in these
metaphors is what is missing in the bourgeois imaginary at large—a nuanced understanding of the historical experience of actual bodies under the
fossil regime and the structural constraints that divide those bodies from
one another through a persistent dialectic of upper-class (and, to some extent, middle-class) privilege and subaltern suffering. While the language of
the mechanical slave and the energy slave is powerful in its ability to evoke
the *amount* of disembodied labor we collectively rely on today, and it helps
to dramatize how *some* bodies have been emancipated by this development,
one unintended consequence is that it blinds us to the persistence of the
muscular economy that still dominates much of the globe where bodies are

used up in old ways and where class exploitation has only been intensified and economized through the application of carbon's power. The problem is that the metaphor of mechanical slavery has been deployed up to this point in a way that steers the mind away from the persistence of physical labor in the current fossil economy while stripping the terms *slavery, servitude,* and *exploitation* of their ability to connote socially specific types of suffering. To paraphrase the anthropologist Alf Hornborg, fossil fuels did not *replace* human labor; rather, they *displaced* it by rendering physical exploitation less visible to the privileged.[50]

That is to say, our language and our politics carry the social landscape in them, and the body of Rastus calls us back to that fact. It speaks, for instance, to the problematics of carbon and class. The trope of technological servitude has always been an unintentional class-based metaphor, it has always been the language of the bourgeoisie, directed at those who command that energy and one that has not sought its audience among, or derived its perspective from, those who have found themselves in the path of carbon. Likewise, the body of Rastus as an African American body brings us back to the relationship of carbon and race. It reminds us that mechanical labor has worked along racially stratified lines—that the virtual slave did not end racial exploitation but rather introduced into the world new racially contoured geographies of pain where minorities across the United States and global South were too often excluded from the privileges of carbon. And finally, Rastus's body reminds us that the fossil economy is gendered—that while labor has often been pictured as a male burden, women's labor has always been deeply woven into the warp and woof of history.

Rastus's body—artificial though it might be—brings us back, that is, to these socially differentiated bodies and their place in our history and our future. It reminds us that it is too easy to speak in abstractions—of averages and medians—when we leave behind actual bodies. To confront climate change we will have to bring these bodies back to the table and take inventory of their values, their pains, and their fantasies for a postcarbon future.

Second, the body of Rastus speaks to a collective failure to understand how nature has been administered under the terms of the fossil economy and, in particular, under fossil capitalism.

From the advent of the steam engine, the mechanical slave has implied the circumvention of nature—an escape from the old rules where labor meant bodies, where bodies meant food, and where food meant thinking carefully about soil, trees, and other limited natural resources. The energy slave allowed some of us to forget the logics of what we call the organic economy and to imagine that technology had somehow detached us from the environment. The phantasmagoric spike in wealth that occurred in the industrial core in the wake of these carbon technologies gave many of us, that is, the privilege of being very imprecise in our thinking about natural resources, technology, and wealth.

This imprecision in thought is carried forward in the literature of the energy slave itself. The current method used to derive estimates is to take Department of Energy data of total energy consumption and divide that number by the work potential of an average adult male estimated at seventy-five kilowatts of power. What is lost in translation is how that energy is actually utilized, how it flows through our lives, and who owns and controls it.

One part of the problem is that this trope simply misrepresents the ecological role that carbon technologies have played in modern life. Fossil fuels do much more than perform physical work. They free up arable land by taking pressure off forests previously needed to provide fuel; they relieve soils of the former caloric demands of biological labor; they provide the surplus heat needed to construct a world of congealed carbons, of concrete, steel, glass, titanium bikes, and metallic solar panels; they ground our food supply; and they provide the main material input for the production of synthetic fibers, rubbers, epoxies, plastics, laminates, and fertilizers now critical to modern culture.[51] Hard and soft coals, short and long chains of hydrocarbons, enter into our lives through different processes and for different purposes, and there are thus different reasons for mining anthracite in Pennsylvania and bituminous in Wyoming's Powder River Basin.

Energy, work, and technology might bear a close kinship to one another, but they are not perfectly synonymous, and thus part of the political work that needs to be done requires reckoning with the multifaceted ways that fossil fuels structure how we sit, at seven billion people, relative to nature.

That reckoning is, however, also a political one. The potentially bigger problem, as Bertrand de Jouvenel once said, is that our command of nature's energies and our access to technology do not fall under our management in the way that these terms "servitude" and "slavery" imply.[52] These energy slaves, and the industrial processes of modern life they reproduce, are not really owned and commanded by individuals or families but rather accessed in a premapped infrastructure of highly centralized corporate power that structures how we gain access to them and how we don't. We know that nature's bounty is not collectively shared, but it is important to remind ourselves that those who hold and pull carbon's commodity chains occupy a privileged position in administering our relationship to nature, whether we are middle class, working class, or poor. What our relationship to nature is today and what it will be in the future requires, that is, a frank confrontation with the structures of economic power in the world that have for too long made these decisions without our input.

And third, that body calls us back to the erasure of historical memory in the politics of economic development and prognostication.

It reminds us that neither past nor current climate politics lend themselves to neat liberal, neoliberal, or progressive metanarratives evacuated of historical difference. The conceit of inventing a mechanical man to free humanity from the material world was always a premade parable rather than a tested hypothesis or historical fact—a fable that has for two centuries insisted that we might emancipate ourselves through reason, science, technology, and the mastery of nature. That enduring technofundamentalist fantasy has been well suited to free market capitalism, with its emphasis on technology, private property, and the liberating effects of human creativity, and it has proven equally adaptable to political ideologies on the left and right. But today the danger is that this language of servitude reinforces neoliberalism's retrograde climate politics by emphasizing mastery and by functioning as the metaphor and evidence that gives to its story of progress a dose of credibility while also making its ideology legible.

Today's climate activists, of course, understand that the relationship to fossil fuels has never been suitable to a story of emancipation, that our carbon technologies have always evinced a Janus-faced history that doles out wealth and penury, pleasure and pain, comfort and risk in equal measure, but this rhetoric of energy slaves risks leaving us, in Pritchard's words, "dangling in abstraction," as our politics concern actual people.[53]

To unthink today's carbon culture means unthinking overly generalized language. It means digging deep into the modern soul to bracket its reason and common sense, these historically specific things we inherited from a unique, and unrepeatable, moment of great ecological generosity, and it means reckoning with the fact that we are an odd type of self-rationalizing animal with some hard-wired genetic and cultural coding that coaxes us on toward a smoke-grey horizon in spite of ourselves. If we are to rediscipline the modern soul for a postcarbon future then those of us who profit from this energy regime will need to distance ourselves both from abstracted metaphors of empowerment (that repress the costs of that power) and from the common sense that the technologist's fantasy perpetuates. A sustainable future will mean rethinking the tropes of nature as servant to humanity and humanity as master of nature in order to bring our technologies back into line with the type of environmental and humanistic framework that comports not only with our understanding of the appropriate use of energy but also with our sense of social justice. Today's political project will require, that is, reweighing and rewriting the history of the industrial age and the postindustrial hangover by confronting not only its successes but all of those blocked desires, those broken bodies, and those social failures locked up in a history that we have until now collectively repressed and struggled to articulate.

CO_2 levels have climbed upward of four hundred parts per million. We face the prospect of entering a hothouse not seen since the time of the megatooth shark. To be ready for that future, we will need not only new technologies and new social policies but also the right language and history.

Four

Fossilized Mobility

A Phenomenology of the Modern Road
(with Lewis and Clark)

The road precedes us. It is a priori, and when joined to combusted carbons —to steam, diesel, and combustion-engines—it becomes assumed: background rather than character, thing, or event. Another way to say this is that concrete and asphalt—the products and pathways of carbon—give us the misimpression that movement across space is effortless, secure, and uniform, that it is ahistorical, placeless, and flat, that it is, in short, an interstice, or space in between.

Today we move without the same degree of cognizance of physical connections between spaces that walking by foot or riding by chaise cart invited and thus without experiencing the same degree of physicality of the road, the cobblestone path, and the leaf-cluttered trail. We leave here, we show up there, and in between is a perceptual blur, a spacelessness, or, at best, a detached landscape seen through glass. We move from "interiority to interiority," Rebecca Solnit writes, or as critic Stephen Kern puts it, we now inhabit a different "culture of time and space" than did our predecessors, one where the sway of grasslands, the heat of concrete, and the carving of riverscapes appear, if they do appear at all, as disembodied and at a distance.[1]

It is only through great effort that we can unlearn or denaturalize the fossil economy's culture of time and space, look outside of ourselves and imagine the body moving, feeling, and intuiting differently than it does. We will need to do that, however, if we are to move conscientiously past this ecological crossroads.

The individual subject of automobility. *Carol M. Highsmith/Carol M. Highsmith Archive, Library of Congress*

The Road in Historical Perspective

American lives are predicated on an infrastructure of carbon mobility. The critic Cotten Seiler has argued that the practice of automobility is, in fact, a core apparatus for the construction of modern individualism in the West and its tropes of freedom and social mobility.[2] (It is certainly a fixture in fossil capitalism's appeal to, and seduction of, the masses.) This personal mobility is something like a social contract today, and it is backed by a transportation infrastructure that is now transcontinental and global in scope —linked together by personal automobiles, transoceanic jets, high-speed rails, and four-hundred-million-pound container ships. If we limit our consideration of the scope of this infrastructure to the United States, then 75% of all petroleum consumption and 28% of all energy consumption go into transportation alone. Americans collectively now drive something like three trillion miles a year, or what amounts to the entire distance from Earth to the planet Pluto, round trip, 337 times over. Globally, more than 60% of global petroleum consumption and 25% of energy consumption in total go directly into transporting bodies, resources, and goods, making automobility a key feature of modern life in the West and a key contributor to climate change.[3] The energy deepening implied in this movement and its attendant transportation hardware make it possible to have a job in Florida and a family in San Diego, to depart Los Angeles on a sunny morning and to arrive, without the body having experienced the cold muscular vastness of the Plains, in Detroit, nearly three thousand miles away, on a snowy evening. This style of living is shocking, to say the least. But for much of the West it has become jejune, second nature.[4]

The road was not always assumed. Moving bodies and goods across time and space was a fully embodied act for most of human history. In fact, for our grandparents and great grandparents the road was often more of a question mark than a fact. That was true well into the twentieth century. As late as 1919 Americans did not know with much certainty what the state of their own roads was beyond the orbit of home, and it was thus difficult to predict what they might encounter around the bend. In an extraordinary response to this uncertainty, Colonel Dwight D. Eisenhower, who would later authorize the construction of 41,000 miles of interstate highways, led a federal expedition from Washington, DC, to San Francisco, CA, to ascertain the state of repair and presence (or absence) of local and transcontinental roadways. Departing from Landmark Zero on the East Coast with a busy caravan of eighty heavy- and light-duty trucks, personal observation cars, motorcycles, a tractor, ambulances, kitchen trailers, and 280 men, his convoy crept along for two months at six miles an hour on poor or nonexistent roads. At times the convoy slowed down to thirteen miles a day, or slower than the pace of a healthy walker. The state of the nation's roads was, it turned out, primitive and frequently unpassable. From Illinois to California the convoy ran on dirt, on sand, over mountain trails, and through mud, dust, and fatigue. Reports and images show men dragging trucks through long stretches of "gumbo mud" in the central states, pulling them out of "quicksand" in the far West, and constructing, by hand, "wheel paths of timber, canvas, sage brush or grass for long distances" to make the going a little less rough.[5]

Today that trip takes a long three days, and the main risk is falling asleep at the wheel. But for Eisenhower it took sixty-two days, 250 accidents, and twenty-one casualties to make it across those 3,281 miles, even with the assistance of oil-fueled motorcycles, cars, and trucks. Moreover, that mobility required considerable human energy, or muscular exertion. Reports tell us that the convoy arrived on the West Coast having survived on five and a half hours of sleep a night and an "excessive amount of strenuous work, . . . lack of shelter, ration difficulties, lack of bathing facilities, and at times the scarcity of water."[6] The point is that prior to the postwar era of the automobile, before the 1954 Interstate Highway Act and before commercial domestic flights, the modern symmetries of time, space, and the body were still being worked out on the continent. Movement, beyond familiar routes, was still both an unpredictable thing and a physical, muscular, and registered thing, known in the joints, muscles, and bones (as it remains for much of the developed world). The road could not be taken for granted.

To gain perspective on the nature of our own lives, to grasp what it was like to move before automobility—before the high-speed rail, the intercontinental jet, and the horseless carriage—we might spend some time in the logbooks of Meriwether Lewis and William Clark, these two archetypal trekkers who moved across the plains, rivers, and mountains of the continent, sometimes on foot, frequently on horse, and mostly on the nation's waterways, with a rare conscious regard for recording the phenomenology of the road before prehistoric carbon.

The Phenomenology of Lewis and Clark

Movement has a history, and there is a phenomenology to it as well. Wind in your hair from an open-top convertible, tired shoulders from clenching to the neck of a horse, the roll of the sole on a pebbled wash, the hum of turbines on an Atlantic flight—each entails a different embodiment. When Lewis and Clark set out from the East Coast in 1803, they knew this, knew that the road was a sensuous experience. In fact, Lewis scribbled an undated note to himself, sometime along the way, defining the word "sense." He explained that "sense" is "a faculty of the Soul," and that it concerns how external objects make "impressions" on the apertures through which our bodies see, hear, smell, feel, and taste "the Ideas" of the world around us, its "different forms, hardness, or Softness."[7] We breathe in the world, he thought, and that breath informs our knowledge of it as well as of ourselves.

So we might ask, what did these archetypal trekkers see, feel, hear, touch, and taste as they moved across the continent? What did being on the road, off the road, and on the river feel like to them? Was there a phenomenology to the road before asphalt, and if so, might it tell us something about—or provide some historical context for understanding—our own habits, substance, and expectations of mobility?

Some context is needed. These travelers were not, as I am today, looking for Lolo's hot springs nor were they hoping for an hour of fly fishing after a day of work. They had been commissioned by then president Thomas Jefferson with finding the easiest navigable route to the Pacific, the aim being to improve commerce with the West's indigenous tribes and to map out the prospects for the nation's expansion into recently "purchased" territories. They were, in other words, the voice and body of empire, and in that capacity they had been deputized to keep their eyes peeled and their ears to the ground, to pay special attention to the potential for road laying, and to work as quasi naturalists, geologists, and anthropologists by recording the flora, fauna, minerals, and peoples that inhabited the path.[8] Moreover, they were not the first to travel down this road. Various Native American tribes knew this geography and these paths well.

That is to say, the journey of Lewis and Clark had a certain sobriety and implicit violence to it that contrasts with the dharma-bum quality or tedium of the postwar road. Yet the record they left behind is unparalleled. Day to day documentations by Lewis, Clark and several of the other men accompanying them provide us a rich phenomenological record of what it felt like to be on the nation's first official cross-country road trip.

The Embodiment of Nature: "Impressions on Certain Organs of the Body"

The first thing that stands out in the phenomenology of the pre-asphalt road is the *embodiment of nature*—the body's saturation in the natural world and the immediacy of its apertures to the materiality of that world. Today, the car, rail, and air travel are forms of insulation against the environment. We are tied up in "artificial corridors," William Cronon has said, where the body and its relationship to nature's variability are no longer front and center.[9] Although we still slow down at the tin sound of rain pattering on a car rooftop in recognition of the increased precarity of the road, modern automobility otherwise keeps the natural world at bay, except when the car's heat is not working or the tire has gone flat. Arguably, that insularity produces something like a "derealization" of the self, a self that sits "abstracted" from its body and the world it encounters diminishing the embodied knowledge we have of the world.[10]

This insularity of the body from the road, from its elements and its textures, is among the reasons we struggle today to understand our relationship to the environment we walk through. Without such embodied knowledge, it is harder to really know the topography of a place, the composition of its dirt and soil, the transition of its plants through the seasons, the workings of its weather and water in supporting life, and the role of the human in that environment. The consequence is that what was once commonplace knowledge is lost to everyone but the expert.

That ignorance in the face of climate change inches us further along toward the precipice.

What first strikes one in the journals of Lewis and Clark is the immediacy of the sensuous world and the porousness of the premodern body to that world. The presence of sight, touch, smell, hearing, and even taste in the men's logbooks is palpable.

For example, sight was then, as it is now, a privileged sense, and it dominated the consciousness of mobility. Every day was a cataloging of the sights: "immence herds of buffaloe Elk deer & Antelopes," "high hills," "conical mountains," "prickly pear," "white geese," "swans," "salmon," and "crayfish."[11] Being close to the ground, and moving at a walker's or floater's pace, the ecology of the region was more visible to the men than it is to us today. It was easier for them to see the workings of the plant and the fish as well as the field and stream—and thus to have a fully embodied experience of the natural world. Lewis wrote regularly about the species and lifecycles of the flora and fauna he encountered along the nation's rivers, prairies, and mountain passes. In eastern Idaho, he saw, for instance, a "singular" plant in bloom, what we call ragged robin, which he noted "grows on the steep sides of the fertile hills [its] radix . . . fibrous, not much branched, annual, woody, white and nearly smooth." Day in and day out we hear observations about nature's life cycles: "the wild rose appears now to be in full bloom," "the salmon are plenty . . . but dying," "the thissels . . . high now in blossom." These were all part of a more general empiricism that grounded this transcontinental road trip, and they make clear that it was easier to maintain the eye of the naturalist at a walking speed.[12]

Similarly, the tactile, the touch of things, defined the men's experience of the premodern road. The body knew the road intimately, felt it differently than does the modern driver behind the wheel pushing a pedal. Most of that feeling went unrecorded, but not always. For example, while walking on the western prairies, Lewis wrote that the men felt the earth through thick leather "mockersons with double Soles" as they walked over fields of "prickley pear," over a stretch of prairie "so hard as to hurt the feet very much." At such times, we can almost feel the crunch and bruising of the road. At other times, leather moccasins were the only protection available as men plunged step by step over mountain passes through snow and ice, their clothes and bodies getting heavier and wetter each foot of the way.[13] Rain-drenched clothing, frozen hands, sun-burned arms—the touch of the road never left them, and such contact with the ground, with the sun, wind, and rain, was incentive to understanding nature's patterns. The men speculated on the meaning of clouds, on when the next dust storm would occur, on what was in store for the mountain pass up ahead.[14] The touch of the road was a way of knowing nature, and it cultivated critical thought about it.

The case is not so different with smelling. As strange as it might sound, Lewis and Clark smelled their way across the continent too, remarking on this sense with some frequency. Most often, it was simply the flora of the region that impressed itself on this sense. The men regularly commented, for instance, on the smell of flowers in bloom and of native fruits that had ripened. Lewis wrote of one cheerful afternoon when they came across a native honeysuckle along the riverbanks in Iowa and how that smell evoked memories of home, since it "smelled precisely like the English Honeysuccle so much admired in our gardens."[15] This sense of smell also informed, however, the experience of the foods they ate and the people they consorted with to get across country. The scent of "aromatic herbs" wafted up, at times, from the river banks, the "horrid smell" of rotting wolf carcasses and "dead fish" hung in the air, the rankness of castor oil from the glands of hunted beaver stung their noses, and the comforting smell of "rosted" roots and "smoked" or "jerked" meat indicated that food and communion was near.[16] Such smells mostly registered outside of consciousness, going unrecorded and sinking instead into the deeper registers of the body. But the phenomenology of the premodern road always had a smell to it.

The ear, in contrast, had a clear value and articulated function. It was the critical sense for giving the expedition a premonition of the unseen risks and uncertainties coming down the road. For instance, just beyond the Cascades on the Columbia River in Washington, the crew became aware of "the roreing of the grand Shutes below." Clark scouted ahead and prepared to take out the boats for portage before they reached what were risky and impassable rapids.[17] Not far from there, they also woke up one morning to the sound of "hard claps of thunder" and the river's "waves tremendious brakeing with great fury against the rocks and trees." That signaled to them that their situation was "dangerous," that waters were rising and that they needed to move quickly and relocate their riverside camp from the banks to a nearby marshy wetland.[18] More generally, the men kept ears open for indications of human dangers and opportunities. At different times, they heard "Indians singing" or rifle "shots" and "alarms" that helped them to distinguish between friendship and hostility.[19] Of course, much of what the men listened to was simply conversation, to the commands of men to other men and the negotiations between them and periodically to guidance from and conversations with women. The road was after all a human space—a place of hierarchical commands being barked to soldiers, of grumbling when supervisors were not around, and of laughter, singing, and conversation throughout long dull stretches of the road.

Finally, the fifth sense, taste, came into play only at certain times; its value on the road was primarily that it enabled the men to gauge the quality of uncertain provisions in an unknown ecology. Taste buds were important for deciding which native roots, herbs, and fruits were palatable and which were not, which were "agreeable" or "disagreeable," and which made one "vomit" or gave men pleasure. In winter high up in the Rocky Mountains, for instance, the men were forced to live on spoiled elk, which had become "verry disagreeable both to the taste & Smell," and it was only the availability of roasted roots that helped to offset that disagreeableness.[20] But the importance of taste, of the tongue, was absorbed into the larger ecology of fueling human prime movers to keep them healthy, compliant, and on task.

In sum, the pre-asphalt road demanded a full-bodied engagement with the ecology, topography, and texture of the natural world. While it is too simple to draw a black-and-white contrast between the old road and the modern road, it is hard not to see that today's automobility encourages a disengagement of that body from nature, a disengagement produced by the increased distance of the mobile body from the world's flora, its dust, its rain, and its seasons. Not knowing the road through the senses in this older way carries with it certain liabilities. Climate-controlled interiorities insulate us from the natural world we rely on, and that insulation means that modern heat waves in California, buckets of rain in Maine, these little early signals of climate change are less visceral to us—less meaningful when we don't walk through them. They don't register until they have achieved size.

The Temporality (or Irrationality) of Movement: "We Could Not Move Today"

The second feature in the phenomenology of the premodern road concerned its unique *temporality of movement*. Much has been said about how railroads, automobiles, and wireless communications have collapsed time and space, allowing us to exercise an unprecedented degree of control over our body's relationship to these two coordinates. The cultural historian Stephen Kern has famously argued that the increased rate at which we move, the perceived acceleration of time we inhabit, and the lived experience of being immersed in this high-energy world—from the feeling of the daily commute to the modernist aesthetics of ragtime and George Braque—are products of the rationalizing of time and space, the recalibrating of them into standardized units that allow for a more predictable rate of exchange of the body and information across distances.[21]

This precision we have achieved in determining how the body moves through time and space is part of today's social contract, an agreement written into our tax dollars for subways and buses and highway maintenance. We feel affronted when we are not given advance notice that the highway is to be closed for repair, since it means a cascade of missed appointments. And we map out our aerial travel across hundreds of miles within forty-five minute layovers, based on the large presumption that planes will run across oceans and continents at the very close intervals to which we have set them to run. This rationalizing of the road is deep in our culture, and it defines our consciousness of temporality.

In contrast, Lewis and Clark did not think in such measured units. They were the agents, rather than subjects, of inscription, charged with rendering the road more legible, more predictable in the future.[22] The original estimate (made by President Jefferson) was that the men's round trip would take one year. It took a staggering two years, four months, and ten days. Not minutes, not hours, but instead years characterized the difference between expectation and experience on the premodern road.

For Lewis and Clark, a particularly good day entailed steering downstream on a strong tailwind and riding a quick current in cooperative weather. Such a day amounted to a good seventy miles traveled, or what today we might cover in an hour or so of driving. But such speed was a rarity: the distances covered from day to day varied considerably.

> September 5, 1803. "Again foggey[;] . . . had some difficulty in passing several riffles. . . . Came 16 miles today."[23]

> May 4, 1805. "The country on both sides of the Missouri continues to be open level fertile and beautifull as far as the eye can reach. . . . Came 22 miles today."[24]

> July 11, 1805. "The wind verry high from the N. W. which oblidged us to lay at Camp untill late in the afternoon. . . . [W]e floated about 8 miles [today]."[25]

> March 17, 1806. "Shall Set out as Soon as the weather will permit. . . . [W]e fear by waiting untill the first of April that we might be detained Several days longer."[26]

Sixteen miles. Twenty-two miles. Eight miles. Zero miles. The pace of movement across country was unpredictable in a way that it rarely is today. Over the course of the entire road trip, Lewis and Clark could claim only an average of *six miles a day*, or half of a mile an hour if we assume a long twelve-hour day of travel. That pace was not much quicker than the pace of an aimless walker. This is because many days—sometimes months at a time—the expedition could not move at all despite men's desires. On these days, which were figured into the costs of movement, the forward progress of the expedition added up to zilch. Such days included long winter months when the men sat at Fort Clatsop in Oregon waiting for the road to melt and reopen; diplomatic days when bartering and negotiating the material (and political) requirements of passage meant smoking, dancing, and giving gifts to friendly tribes to acquire work horses or local knowledge; and intermittent days of fatigue when a particularly harsh part of the road required some recovery time for living prime movers.

Averages, in other words, were not very meaningful in the phenomenology of the premodern road. The data points were simply all over the map. Patience, irregularity, and unpredictability trumped the metronomic quality of the modern highway.

A second feature of the temporality of the premodern road was the resting required to accommodate the limits of human and animal endurance. Today on a full tank of gas I can expect to travel approximately four hundred miles before the next refueling. Assuming that I have a healthy bladder and a full stomach, I can furthermore anticipate traveling about six hours at a stretch before breaking for thirty minutes at a rest stop. Lewis and Clark, in contrast, had to punctuate their movement over much shorter periods of exertion to accommodate the limits of the human motor. Throughout the trip, we hear Lewis and Clark commenting that the men are again "fatigued" and require "refreshing."[27] Private Whitehouse, for example, noted that although February 22, 1805, was a pleasant day, "the party rested themselves, still being fatigued" from having the day before "halled a heavy load 21 miles on the hard Ice & Snow."[28] During these slow times, labor did not stop but forward progress did. Men still had to hunt out suitable spots along the road to recuperate, to prepare food to recharge men's batteries, and to ensure that horses were well oiled and gassed. If there was any hurry up and eat, that went unrecorded. Without rest, without food, and without repairing tired muscles and fatigued brains, no one was going anywhere.

A third feature that should be fairly obvious at this point is that weather, topography, and ecology dictated the relationship between time and space on the premodern road in a way that it only occasionally does today. Although some mountain passes still close in the winter time, because too much work is required to keep those roads open through heavy snows and although heavy flooding, blizzards, and hurricanes periodically shut down main arteries along continental coasts and riverways for a few days every year, the norm—the expectation of the modern road—is that the work of transcontinental mobility can, and should, go on unimpeded both day and night and across the seasons. The expectation of such control over the environment was foreign to Lewis and Clark.

Rainstorms, suffocating heat, steep mountain inclines, and insects—the environment imposed itself on movement over each step of the preindustrial road. Rain, for instance, brought travel to a standstill or slowed it down considerably many times over the two and a third years of travel. For example, one morning near what would be their winter quarters at Fort Clatsop in Oregon a storm raised the tide of the Columbia as much as two feet while the men were still in bed. That rain flooded the men's camp and set everyone to work moving goods and simply trying to stay warm and dry. "All wet," Clark concluded, and his sergeant pointed out the obvious: no one was "able to set out" on the road that day.[29] Wet robes and cold bedding were a common complaint across the years, but its opposite, the heat of summer, shaped the temporality of the road too. On the border of Kansas and Missouri, for instance, the intensity of a July sun on the plains, Clark explained, forced the party to delay "three hours to refresh the men who were verry much over powered with the heat."[30] At another point Clark wrote that "being hot the men becom verry feeble."[31] If the weather was a determinant, so too was topography, which was also felt every step of the way. It was one thing to row upstream; it was another to sail downstream. It was one thing to load packhorses for a journey over the Rockies; it was quite another to float down the Ohio. For instance, towing a boat upstream against "a current running very strong against us" with a summer sun beating down could make mobility unbearable: "It can hardly be imagined," Sergeant Whitehouse wrote, "the fataigue that we underwent."[32]

But the environment could impose itself on the phenomenology of the road in more insidious ways. For example, without automotive power and the protection that plate-glass windows provide, even small pests could wreak havoc on a traveler's best-laid plans. As the expedition neared the end of its journey, the mosquitoes along the river became so bad that grown men had to "retreat" from them; the pests, they complained, were "so noumerous and tormenting" that it was "impossible" to labor that day. That particular day must have been awful, as Sacajawea watched while her child's face "puffed up & Swelled" and as everyone in the crew simply struggled to make it to nightfall, hoping that a breeze might pick up to whisk pests away. Snow, rain, heat, sand, insects, steep inclines, white water—the humility of working with and against nature has few modern parallels. This sensibility acquired from knowing nature is lost on the hurried today.

Finally, the more measured temporality of the preindustrial road was also defined by its nonlinearity, unlike the modern highway and air route. Today, carbon technologies have flattened the highway and straightened the road out to a considerable degree. We have carved out a fairly straight beeline across the continent, cutting through the Alleghenies rather than going over them, driving over the Mississippi River rather than fording through it, and moderating the hills and dips along the way. By contrast, Lewis and Clark followed a meandering path across the continent. The primary reason is that they were forced to travel by water whenever possible, since floating and rowing—which reduced friction—were less laborious than dragging bodies and goods over land. They thus were drawn to follow the path that the continent's rivers had cut through the plains, and that path was decidedly circuitous. The riverine route meant traveling downstream south and then west and then southwest on the Ohio River, traveling north on the Mississippi River, and then meandering northwesterly against the Missouri River up to the foot of the Rockies. Faced with a wall of mountains, the crew then buried its boats and some goods, traded local Indians for horses, and rode west over those mountains before picking up the Columbia down to the Pacific.

The logic of the river has always been different from that of the highway. It does not concern itself with the efficiency of distance but rather with gravity's path of least resistance. The river searches out the easiest way to the sea, it follows the topography of the land that gives to the riverine road its unique meandering character. Sometimes one traveled backward or sideways to go forward, and thus there was no measuring progress as the crow flies. Lewis took coordinates daily and he mapped out the distance traveled, but it is hard to see just how complex his movement was across the map. Rather the nonlinearity of premodern mobility is best seen in the sketches he made of the road. In the narrows of the Columbia, for instance, Lewis drew a river that was sinuous, slithering like a snake toward the Pacific. Minimizing labor rather than distance drove the logic of the premodern road, and so scouts and men plodded westward, northward, southward, and eastward step by step.

Automobility has changed that logic. Disembodied labor, a surplus of power, allows us to focus on the time it takes to cover distance rather than the labor involved in getting us from here to there. The logic of the asphalt highway exchanges the meandering path for the thoroughfare. But in doing so, it distances us from the topography and geography of the continent, and that means a little lost knowledge about the world that houses us.

The Muscularity of Movement:
"Extreemly Hungary and Much Fatigued"

The third feature that stands out in the embodiment of the preindustrial road is the intense *muscularity of movement*. Lewis and Clark, to paraphrase the historian Richard White, knew nature through labor. Their journals remind us—in fact they insist on it time and time again—that movement means work, that hauling two six-foot-tall, well-built bodies and an expeditionary force of forty men and their belongings across any distance required considerable energy resources and a substantial quantity of coordinated and individual muscular exertion.

Movement today is, by contrast, mostly disembodied. There is no carrying myself over Lolo Pass unless I choose to hike with my backpack. Rather I *am carried* by little charges of combustion converted into forward movement that send me hurling uphill. That disembodied labor makes the self feel as though it extends beyond the corporeal body, and it unconsciously feeds our psychic attachments to fossil fuels.[33] But because this labor subsidy is hidden beneath the hood of the car, the relationship between work and mobility can easily pass us by. For instance, the term *horsepower*, which is meant to invoke the labor involved in movement, only vaguely evokes the depths of our mechanical labor dependencies. The mind trips over itself trying to imagine 175 horses chained to a steel chassis pulling a car over a hill to the whip of a mule driver. It has further trouble imagining that long train of work animals reaching the top of a hill depleted of energy, taking up space on the grass, and looking around for a large meal of oats or foliage to refuel. The term, in other words, fails to do justice to the uncanny reality that burning ancient plant matter permits us to do an end run around the labor and resources once required to hurl bodies through time and space.[34]

Lewis and Clark commented on this fully embodied relationship between muscularity and mobility with frequency.

First, they noted that moving anything of weight, such as a sack of food, meant heavy work, labored breathing, and the need to refresh bodies at regular intervals. For instance, outside of Pittsburgh, Lewis and his crew (sans Clark, who had not yet joined the group) encountered a portion of the Ohio River that was too shallow to row or sail on. In the face of this riffle, the near-frictionless demands of sailing gave way to a muscular heaving over the river floor. Getting over this stretch of the Ohio required calling "all hands to labour" for two hours of heavy exertion before the men's fifty-foot keelboat and goods could be dragged over the patch and put into water deep enough to allow the vessel to float again. A whole lot of work and sweat produced, in this case, a very short distance traveled. Lewis was brief in his notes: "We passed [this] little horsetale ripple or riffle with much deficulty." But the rich meaning of the phrases "with much difficulty" and "all hands to laboring" should not be lost on us. They are meant to invoke the physicality of a world of work wherein the so-called hand—this metonym for the working body as a whole, its lungs, heart, its opposable thumb, muscular back, legs, and torso—once grabbed and wrestled with the natural world (and against itself) often with "much deficulty" in order to effect any kind of real or figurative forward movement.[35]

Second, the somatic character of premodern mobility found its organic limits in the human body. Few people, excepting the solitary mountain man and the individual scout, moved alone across the continent on foot for much distance. Traversing any long distance meant being heavily weighted down with food, tools, fuel, boats, tents, packs, pets, guns, blankets, and beds, among other things. And thus mobility was an exercise in association, tied to the help of family, friends, strangers, servants, or conscripts. Although Lewis and Clark appear in our history as heroic individuals, their names stand in for a larger supra-organic body made up of the coordinated and disciplined energies of other unacknowledged and barely recorded bodies. In fact, much of what these men did was manage other people's muscles through discipline and incentive. The carrot in the equation included the promise, made on the part of the federal government, that each conscript would receive a soldier's section of land upon return. That was a way of offering future social mobility for immediate hard labor. On a more intimate level, incentivizing coopera- tion meant doling out small rewards to the men. Lewis and Clark regularly permitted the men to "refresh" themselves: they frequently gave "the men a drink of sperits" to soften the hard edges of the day, and they regularly allowed for men to spend the night "in high Spirits fiddleing," or dancing, as a type of reward for their work.[36] The stick in that equation included the threat of discipline and its occasional public use. For instance, the trial of Moses Reed on August 18, 1804, for attempting to desert the expedition ended in him having "to run the Ganelet four times thro," enduring an un- known number of "lashes," and being excommunicated from the "Permonant Party."[37] Of course, certain persons, like Clark's African American slave, York, experienced these instruments of governmentality in a particular way. That working body did not simply go along for the ride but was coerced by a racially determined somatic labor regime that threatened the life and limbs of people the state had designated to be property. The heroic journey de- pended, that is, on the collective rather than the individual, on cooperation and discipline to extract a nonvoluntary quotient of labor. There was no bourgeois individualism on the premodern road.

Lastly, the muscularity of the premodern road required, as mobility still does today, technology to leverage the hand of man. The logbooks of Lewis and Clark remind us that the human body, even when amassed in numbers, produces insufficient force to overcome gravity and inertia when the going gets tough. There was, for instance, no lugging a keelboat, pirogues, and men over the mountains, through rock-laden white water, or over a timber-filled river without prosthetic devices and energy converters that extended the human hand, such as ropes of horsehide, iron-pointed levers, and bellows and augur planes to make and repair vessels, as well as organic technologies, such as more powerful energy converters like the domesticated horse (and the harnesses that could leverage that force). While still in charted territory on the Ohio, for instance, when human muscles proved inadequate to the job—when men "exert[ed] all our force" but still found the job "impracticable"—Lewis went in search of a larger prime mover. He came back with a trained team of oxen, each of which could pull nearly two thousand pounds of weight over flat land, substantially more than what an individual man could pull. Harnessed together, these oxen could generate the force required to heave and ho a boat over a river bottom. And yet oxen were cumbersome, they required rest, they moved slowly, and they took up substantial food and space. Thus no one planned to bring them along for the rest of the ride. Instead, once the expedition left behind well-settled lands, when the limits to men's muscles were reached west of the Mississippi, Lewis and Clark buried items to pick up on the way home and negotiated with local Indian tribes for horses that might do the job that people could not. Indeed, they fretted about acquiring horses from local tribes on both sides of the Rockies, and they remarked favorably on those tribes that were friendly enough to give them up and regarded with animosity those that were stingy. Traversing the nation's mountain passes was more than the men could handle on their own.

Today, the work required to move bodies, goods, and information is hidden from sight, taking place under the hood or in an electric generator somewhere unseen. Without having to walk and sweat over a steep incline and recover from that work, without having to drive, feed, and groom horses, without knowing the road through labor, those bodies once required to accomplish that work disappeared into insignificance, sank into polyester seats. But there is something lost in our hypermobility, something that cannot be fully retrieved without working the road.

The Sociality of Movement: "The Party Amused Themselves Danceing"

The phenomenology of the preindustrial road had a fourth feature: *the sociality of mobility*. Today, the portability of the combustion engine, combined with the dense energy in petroleum, allows the individual a hyperflexibility of movement. That has proven true especially for the white middle-and upper-class male but even so (if in a more complicated way) for classes, races, and genders written out of other opportunities.[38] Few material obstacles, other than time and money, stand in the way of an individual travelling five hundred miles by oneself in North America over the course of a day or two, and that individualized experience of movement, this "mobile privatization" wherein we experience public spaces in the privacy of our own heads, can be liberating. It can, Seiler says, allow the driver to feel "extrude[ed] from the claims of society."[39]

The preindustrial road, by contrast, required a different kind of association with others. In particular, it required negotiating with people and socializing with them on a near-continual basis. If today the sociality of the road is limited to the conversation of passengers in the car and brief interactions with cashiers at the gas station (a sociality embedded in the hidden structures of taxation and policy decision making), the pre-asphalt road was a more personalized lesson in collaboration, diplomacy, hierarchy, and discipline.

The one thing that Lewis and Clark knew before they departed into uncharted territory was that to get very far they would have to rely on other people, strangers that they did not know. No one thought they would make it to the Pacific without help. In a time before rest stops and fueling stations, they counted on running into indigenous peoples along the road and banked on securing from them needed provisions, horses, and directions. The Corps of Discovery thus traveled across country loaded down with huge sacks of gifts. These gifts, given out according to rank and gender, included chief's coats, hats with plumes, rolls of ribbon, looking glasses, white shirts, medals, hair pipes, wrist bands, arm bands, scarlet leggings, britches, blue blankets, silk handkerchiefs, calico shirts, needles and thread, flags, ivory combs, tomahawks, bead necklaces, tobacco, iron combs, awls, squaw axes, collars of wire, silver rings, broaches, glass ear bobs, hawk bells, mock garnets, Dutch tape, and more. The list goes on and on, and the bulk and weight of those objects was part of the cost of moving across country.[40] For instance, when the second-ranking chief of the Hidatsa visited the expedition in North Dakota, Clark gave him as a sign of friendship "a Medal Some Clothes and wampoms." That night the men danced, as they often did to blow off steam, and at some point some of the men slept with Indian women. The expedition sealed its friendship through speeches and trading while waiting for the

ice in the river to break. Such diplomacy, defined by the practice of gift giving, was part of the social nature of the road that Lewis and Clark depended
on. It guaranteed their safety on this portion of the road, and more importantly, it secured provisions along with Indian knowledge about the tribes
and terrain upstream.

Second, the men counted not only on strangers but also on one another
throughout the journey. For two years and four months, they rowed together,
they ate together, they hunted together, they stood on watch together, they
chopped fuel together, and they talked, danced, fiddled, sewed, and built
canoes together. This intimate cooperation was built into premodern mobility and was necessary to ensure progress and survival in unfamiliar regions
with forbidding obstacles, minimal food sources, potentially hostile Indians,
and a multitude of great uncertainties. The men were, as it were, a corps,
even if the names Lewis and Clark stand in for that collective body, and, as
such, mobility was the function not of individual wills but of coordinated
wills. To facilitate some degree of regularity and security on the road required the cooperation of "all hands," "several hands," or "some men," who
were called to tasks such as cutting the keel boat out of frozen ice or simply
sitting down together to make ropes strong enough to get the expedition
through upcoming rapids. Likewise, when men were injured, sick, broken, or
bruised, others hunted for them, fed them, and ministered to them, taking
up their workload or simply waiting for them to heal. As one example, while
the expedition slept in close quarters during winter at Fort Clatsop, one of
the men came down with a respiratory infection, or "Ploursey," for which
Clark "bled [him] & gave him some Sage tea."[41] In another instance, Clark
prepared a tartar emetic for his slave York when he became "very unwell."[42]
Being on the road was a lesson in solidarity, sometimes willing and sometimes forced.

Of course, a third feature of the sociality of the road was that it was not
egalitarian. It could be relatively equal at certain times and places, but it was
often strictly hierarchical, cut through by status, race, gender, and nation.
In the Corps of Discovery, Lewis and Clark as commanding officers moved
freely, labored less often than others, and had the authority to issue commands. But that freedom of movement depended on the more restricted
movements of other subjects. Clark's servant, York, for instance, moved
and labored like the others but he did so to the commands of his owner. It

is likely he was treated like everyone else whenever "hands" were called to work, but he also found himself beckoned to move in unique ways because of his racial status. For example, on New Year's Day in 1805 while the crew was waiting for the river to melt, Clark "ordered my black Servant to dance" so as to amuse the crowd of Mandan Indians who had gathered for the occasion.[43] A novelty, because of his skin color (and large size), York facilitated the diplomacy of movement in this situation, but not of his own volition. The sociality of the premodern road thus brought with it the degradation of racial classification.

Likewise, class or status mattered on the road. Enlistees, unlike the officers, answered, at least most of the time, to the summons of their superiors when called to the task at hand, and they were subject to discipline whenever they refused to move to the rhythm of the collective body. By not following orders, the men risked lashes, excommunication, confinement, and various less drastic punishments, and they found themselves the object of officers' concerns that they not become "lazy" when they were counted on to move the caravan.[44]

Of course, gender too shaped the sociality of the road. Women, except for Sacajawea, did not experience this particular road in the same manner as men did. Early Anglo-American norms did not accord them the same freedom of movement in public spaces as they did men, nor were women eligible, in this case, to join the expedition, which was pointedly a man's prerogative. Sacajawea, who came onto the expedition in a liminal status—as the wife of a French translator and thus as nonmale, nonwhite, and supplemental to the permanent party—experienced the constraints and opportunities of movement in a way that was unique to her. But her unique role as a racial, gender, and linguistic mediator contributed in its own ways to keeping the party moving forward as they sought out and negotiated with tribes and traveled by the landmarks near her original home in the plains.

Together this complex society of the road in which gift giving, dance, discipline, cooperation, rewards, and collaboration figured largely was a predicate to premodern mobility. While today we still remain a highly networked species, that networking on the road does not entail the same face-to-face contact once required to get somewhere. The power of automobility, the disembodied labor that works for us, promotes a certain degree of individualism and autonomy on the long-distance road that was once known only to the few. A networked but less personal road carries with it, however, its own risks in the form of disinterest in and alienation from those lives that make our movement possible.

The Ecology of Movement: "On this Food I Do Not Feel Strong"

Finally, the pre-asphalt road depended upon an *ecology of movement* that contrasts sharply with the ecology of carbon mobility. To drive from the sun and back multiple times a year, to build a global circuit of information exchange, to shuttle goods in a transnational economy requires gushing flows of energy, and that energy exacts, as we know, substantive ecological costs. Today those costs can be put down to what economists call "externalities:" the environmental cost of drilling, transporting, and refining oil; the resource cost of manufacturing an infrastructure made up of concrete roads, steel automobiles, rubber tires, iron horses, and an aerial fleet; and the cost of greenhouse gas emissions left along the way. In contrast, the premodern road, although it also had heavy ecological costs, left behind a different footprint that tended to be more clearly limited, localized, and relatively transparent.

The first cost of the pre-asphalt road has its analogue in the gas station. Bodies, like cars and trucks, need refueling. The human motor runs on food calories, derived from the soil and aquatic environments rather than from the lithosphere that sits below the soil and seafloor. Food is thus an ecologically different type of energy source from oil or coal, since its production draws on parts of the ecosystem (e.g., soil, rivers, and oceans) that have more obvious limits and that change with the seasons, agricultural practices, and a culture's methods of cooking and storage. But whatever the time, place, and practice, feeding men's muscles on the road was a nonnegotiable feature of the ecology of movement, and Lewis and Clark knew this before embarking west. In preparation, they purchased, quite literally, a boatload of food. The records of one sale of lading included forty-four kegs of pork, seven barrels of salt, one keg of hog's lard, six half barrels of pork, and seven bags of biscuits.[45] Loading, unloading, and moving forty-four kegs of pork plus other dry bulk foods across unpredictable waters and terrain was a constant lesson in logistics and problem solving. But these substantial dried provisions, which likely came from farmlands in the Midwest, were only the initial fuel required for takeoff.

The expedition procured most of its needed calories both by hunting and gathering and by trading with local peoples. To feed forty men or so working hard for ten to twelve hours a day meant constantly negotiating trade deals and laboring to find food. A typical day included, for instance, sending out a small party to hunt bear or net crayfish, trading with local tribes for corn or salmon, drying and salting meat, shelling corn, and cooking meals. This ecological demand of the premodern road exposed the expedition to unique vulnerabilities. For instance, shortly after the expedition crossed the Cascade Mountains, the men found that food calories had become quite scarce. Clark recorded that the hungry party, which had no fortune hunting and gathering, was lucky to run into friendly tribes who though poor themselves were willing to trade "a fiew pounded rotes [roots] fish and Acorns of the white oake."[46] Gaining access to roots, acorn mash, and salmon was critical, and it tied the mobility of the expedition into the broader ecology of the region's indigenous trade network that moved salmon from the river and drew woodland resources, such as acorns and firewood, to it.

Additionally, the party hunted and gathered each day of the trip. The need for protein alone left, for instance, an impressive path of slaughter in the party's wake. Carcasses of bear, wolves, elk, deer, bison, ducks, salmon, speckled gulls, pheasant, mullet, alewives, prairie hens, magpie, and whatever other birds and fish the party could get its hands on by trap or rifle left behind a considerable ecological footprint that made the party's movement possible. At some point, food was valuable enough that the men rejected eastern and tribal norms and started to dine on dogs and horses too (a fact that made them the butt of jokes among tribes who viewed them as lowly dog eaters). Such animal protein was, however, only part of the nutritional intake needed to keep the train moving. The party also counted on gathering fruits, nuts, roots, and various other plants that were in season. For instance, sunburned and hungry, in the heat of late summer, the expedition ran "entirely out of provisions" on the plains. They began, Clark wrote, "subsisting on poppaws" (a tropical-like fruit) that when combined with some remaining "buiskit" proved enough to keep the party going a bit longer, if not happily so. In fact, running out of food, this human fuel, was a continual concern. Even with the considerable help of local tribes, and even with robust mammal, fish, and bird populations to hunt, it was not easy to provide for men's muscles day in and day out along the road. For long periods of time these proverbial work horses went undernourished and sick, were rendered "feeble," as it were. Lewis captured that risk in vivid prose: "My flesh I find is declineing. . . . [O]n this food I do not feel strong."[47]

What we traditionally call fuel (wood, charcoal, and other dried biomass) was also critical to the ecology of early American travel. Such fuel was necessary to stoke fires for cooking raw meat and boiling dried foodstuffs, for keeping men warm at night, and, for, in the case at hand, firing the small iron forge the party carried for repairing metal objects. No one went without fire for long, because travelling over the premodern road exposed a person to the elements and entailed long vulnerable days and nights over an extended period of time. Whenever possible, the party thus simply gathered its own fuel. Frequently along the river that meant harvesting an inferior fuel source like dried willows whenever "there was no other article to make a fire with."[48] But the situation got better when dried wood could be chopped and burned from within striking distance of the road. Still, even in this early phase of western conquest, it was difficult to gather enough fuel in quantity along the road, and so it had to be procured from local tribes who, as Lewis put it, "furnish us with fuel."[49] The scarcity of fuel made for difficulty and delay. In October on the Columbia, for instance, Clark explained that the expedition woke up very cold and "could not Cook brakfast . . . for the want of wood or Something to burn."[50] It tried unsuccessfully to gather enough "dried willows" to generate a fire but failed, and so to make up that deficiency, Clark paid a very "high rate" for precious wood that had to be imported from upriver.[51] This constant quest for something to burn could make the difference between healthy mobility and a risky venture.

A third feature in the ecology of this somatic economy concerned the use of work animals, namely horses and oxen. These prime movers, like men and women, required prodigious energy resources. Oxen were only available in the Anglo-occupied sections of the road and so were used only at the beginning of the journey. But horses, bartered from local tribes, were especially critical to the party once it reached the rugged continental divide separating the Mississippi from the Columbia watersheds. Although sometimes hard to acquire, such horses played a critical role in allowing the party to hunt at a distance from the road, to scout ahead of the main expedition, and to carry provisions (including an iron forge, clothing, tools, tents, and bedding) over mountain passes. Pack and work animals do not, however, work for free. They require, when at work, about twenty-five pounds of grass and other foliage a day and thus need access to pasture or meadow to refuel. Keeping these organic engines healthy meant knowing the lay of the land. Sometimes, as in the mountains, grass was hard to come by, horses became thin, and men risked losing them. These were times when it took good scouting to pick not only the easiest route but the route with the right resources. Along the Lochsa River in Idaho, for instance, foliage was difficult to find. Journal entries indicate that the soil was "indifferent, and verry broken," that the men and horses were "much fatigued," and that the grass had been "entirely eaten out by horses."[52] Other times were simpler times. On Independence Day in 1804 along the Kansas-Missouri border, for instance, Lewis had no difficulty feeding the small number of horses they had at that point: "The Plains of this countrey are covered with a Leek Green Grass, well calculated for the sweetest most norushing hay—interspersed with cops of trees, Spreding ther lofty branchs over Pools Springs or Brooks of fine water."[53] The ecology of movement was more forgiving here. But utilizing horsepower to move down the road, especially when horses were amassed in numbers, meant feeding and refreshing that horsepower a few times each day and leaving sometimes a large swath of grazed trail in its wake. Heavily traveled paths, like the Santa Fe Trail, became over the years exhausted from such heavy travel.

And finally the ecology of mobility depended on a solar energy regime that supplemented and sometimes gave a break to men's and horse's muscles. River currents, or the power of the river, Richard White explains, is part of the work that nature does. Those currents, which derive their energy from the interplay of cycles of evaporation, condensation, and gravity, can be either an obstacle or a friend.[54] Traveling downstream, for instance, amounted to riding on the current when the river was deep enough and when the decline was gradual enough to do so. In central Montana when they were coming home, for example, the men "proceeded on the current Swift passed hills on each side." They made it about nine miles that day.[55] But at times, the topography and grade made the current too tough to rely on and its force overwhelmed the wooden keelboat and handmade pirogues. Such times of "hard water," as Clark put it, could make the going tough. Heading upstream in Missouri, Clark explained that "the Current was so great that the Toe roap Broke, the boat turned Broadside, . . . wheeled & lodged on the bank. . . . [N]nothing Saved her."[56] Too much energy was a problem in this case.

The force of the current was, however, frequently matched by the power of wind. The expedition's main boat came equipped with a square sail that allowed men's muscles periodic rest from rowing and pulling. Although the wind rarely made for easy sailing in a winding river with an unpredictable depth, it sometimes allowed the expedition to "hois[t] sail" and sail "verry fast."[57] Other times, the interplay of wind and current worked at cross-purposes. East of the Rockies in Montana, on the return route, for example, Lewis wrote that "the courant was strong tho' regular" and "the wind was hard and against us," and so the muscular work of men towing by rope was required to proceed through that section of river.[58] Wind and water were only partly predictable, but they mattered greatly to the life of the party, as they were counted on, cursed on, and blessed throughout the years of travel.

To sum up, the ecology of premodern mobility, as exemplified in the Corps of Discovery, insisted on a more clear-sighted reckoning with the ecological imprint of mobility. Fueling organic prime movers meant leaving behind a visible trail of mammal carcasses of deer, dog, and elk; it meant culling the river of salmon, crayfish, and mullet; it meant shooting and trapping magpies and ducks for food; it meant grazing near and along river bottoms, chopping down trees, digging up roots, and pulling up willows; it meant working with and against river and wind; and it meant drawing on the fuel and food resources of Indian communities connected to the road. Such an ecology of mobility tended to put working bodies into a more immediate contact with the ecology of the road and it imposed very different limits on a body's movement.

Conclusions

Today, modern mobility prizes itself on an elastic freedom and predictability of movement across time and space. As one of modernity's most defining mineral rites, automobility sits deep in the affective structures of the modern soul, particularly for the middling and upper classes, who are elementally attached to the fossil economy, but even so for the world's working classes, who are forced into the mobility of migration for work to survive in a frenzied and networked fossil economy. This automobility is nothing less than a structural premise of both globalization and the nation. And it is an underlying assumption of the core geographies of modernity, including the energy-dense city, the decentralized suburb, and the mechanized farm. This movement that feels limitless and has only remote and externalized costs permits the world, for better and worse, to live, dream, and think both in very large and very privatized orbits. It is a foundational part of modern subjectivity—the material basis, as critics have said, of both postwar individualism and the collective imaginaries available to us.[59]

But what is lost in all of this movement?

First, the modern road (this metonym for the weather, climate, soil, and topography through which we move) becomes more abstracted the more insulated we are from it. Being insulated from the natural world is the precondition to a disinterest in it, and it operates as a strong disincentive to understanding the ecological costs of our mineral rites.

Second, this modern road, absent the visibility of sweat and fatigue, minimizes the real work (and energy) entailed in the global transfer of resources, bodies, and goods. It allows us to forget that big orbits do not come free, that in fact the ecological costs in carbon emissions and resource use are nothing short of staggering.

Third, the modern road, once flattened and rationalized, reinforces an epistemology of empire rather than the métis knowledge of experience. It works to amplify the tyranny of the clock, the mile marker, and the abstract over the multiplicity of human experience. This means losing trust in our bodies and our own perceptions and relegating knowledge about the external world to someone else.

Fourth, the modern road, private rather than social, heightens ignorance about (and creates distance between) the divisions among peoples of various ranks, conditions, and races. It makes movement private and semiprivate, encouraging us not to count on others, not to consort with others or see the deep inequalities that are part of our lives, and not to follow out the footprint we leave on others who stand in the path of, and who prop up, our movement.

And finally, the modern road, with its ecological impact reduced to a near-invisible puff of combustion, makes climate change, and our part in it, seem small and distant. Without a trail of chopped trees and hunted deer, without sweating for our fuel, it is that much easier to imagine that there are no costs to globalization, to suburbanization, and to the circulation of bodies and goods.

Today, we have no real choice but to ask what it means to move in the ways that we do, what the priorities of our movement should be, and what nonnegotiable and negotiable attachments we have to our mobility. No one knows what a sustainable road might look like in the future. No one can say with much certainty what scope the road can handle and what values will matter in the future in paving that road. But what would have been clear to anyone two hundred years ago, had they had a reason to think about it, is that the premodern road was not scalable to a continent of nearly six hundred million people traveling down it day after day or to a world climbing to ten billion whose bodies, goods, and resources are jetting about it.

What should be clear today is that every charge of combustion, every forward propulsion, on the road pushes CO_2 a little further beyond four hundred million parts per million. The heat thus comes on a little thicker. We are not so unlike Lewis and Clark in that we only sort of know where we are headed. But what we should know by this point is that we have already traveled too far and that it is time to come home.

Five

Coal TV

The Hyperreal
Mineral Frontier

Oil gets all the attention. Those of us living on the raw edge of the twenty-first century know well that our lives are saturated in petroleum, fingers dipped deep into Gulf Coast wetlands, South American rainforests, and Middle Eastern deserts. The outsized features of the world we inhabit— its unique pace and frenzy, its dense and decentralized patterns of settlement, its rich materiality for the middle and upper classes, and the imperial scope of its politics—are hard to explain without reference to petroleum. The weekly visit to the gas pump, that short sullied point of contact, enacts that dependency in what is in North America a near-universal mineral rite. Geographer Matthew Huber puts it this way: "Oil is now equated with life itself."[1]

By contrast, coal—that other sublimated fossil fuel—is easier to miss. Transmuted into its more benign manifestation as electricity, or *coal by wire*, coal becomes beautified and thus appears to be somewhere else and someone else's problem. When we think of coal as an object, the mind travels back to industrial-age memories of children sorting out rock and slate or of remote Appalachian communities hacking from black lung. Coal tends not to evoke the incandescence of the newest iPhone or the pixilation of a high-definition television. It speaks instead as a historical object or distant artifact rather than a contemporary fact. That is true despite the fact that we burn nearly 30–50% more coal in the United States today than we did when coal was supposedly king and that coal contributes as much as 71% of carbon emissions from electric generation.[2] Coal's *relative* share of the energy budget might have fallen, but we have never really left it behind. And thus how we know coal still matters.

Of course, locally coal remains a very big deal. West Virginia, Pennsylvania, Kentucky, and Wyoming are saturated in the politics and culture of coal. Highways like I-76 cutting through Pennsylvania's mineral-rich country are smattered with reminders that coal mining lives on, that it provides a few decent jobs, and that it is, as one bulletin board reads, "cheap and reliable." Similarly, surveys in West Virginia and Kentucky indicate that people know coal and that they have a conflicted attitude toward it—that they appreciate the handful of jobs that it provides, that they are opposed to its more egregious excesses like mountain top removal, and that they are unclear of the alternative prospects opened up by solar and wind power.[3] But on the national stage, public opinion, which is decidedly mixed on coal, is based on some very large silences and erasures. Americans are likely to know more about China's bleak house of coal-fired power plants than about Wyoming's gigantic Powder River Basin (the latter of which provides 40% of US coal), more about Shell's operations in Nigeria than about the five hundred coal-rich mountain tops that have been removed in Appalachia over the last twenty years.[4]

This misapprehension of the present is especially true in academic culture. Coal does not get the attention that oil does. Humanistic scholars concerned with climate change are, for instance, busily at work building a sophisticated body of literature on our global oil culture. Excellent books with titles like *Lifeblood, Oil Culture, Living Oil, Peak Oil* are showing up on bookstands with increasing frequency, and workshops like the After Oil School in Alberta have been effectively creating an institutional nexus of cultural critics, artists, and various other humanities scholars to make sense of our modern "petroculture." But with the important exception of the ethnological work being done in the coal regions, and a few outstanding books here and there, like Timothy Mitchell's *Carbon Democracy* and Andreas Malm's *Fossil Capital*, there continues to be little interest in this broader "carbon culture" that encompasses coal (and now also natural gas). We are stuck, historian Christopher Jones says, in a sort of "petromyopia."[5] Intellectually, the subject of coal festers; it is cast off as something too base, dreary, and provincial to fit our theoretical fancies. We think we know coal.

Yet coal, like petroleum, needs the voice of the humanities.

What follows is one small pass in that direction. It offers up a cultural critique of a 2011 reality television series that centered on hard rock mining in McDowell County, West Virginia, a part of deep Appalachia that is known both for its poverty and for its one-hundred-year-long history of coal mining. It examines how our knowledge of coal is mediated, managed, how it passes into popular consciousness, and how, in so many little and big ways, fossil capitalism—its acute and generic risk and its startling and enduring inequality—becomes legitimized and rendered, in the words of critic Kenneth Burke, as "natural as breathing." *Coal* (which I refer to as *Coal TV* in order to reify the concept of hyper-real coal) tells us something, in other words, about ourselves—about how we have learned to think and operate under neoliberalism and how fossil capitalism gets culturally managed in the twenty-first century.

The Hyperreal Mineral Frontier

Coal TV first aired in 2011. Brainchild of Spike TV's Thom Beers, who also gave the world *Ice Road Truckers* and *The Deadliest Catch*, it was a popular reality television series centered on hard rock (or underground) mining in Appalachia. The show featured two operators and a small group of hardworking coal miners from McDowell County, West Virginia, who were mining a high-grade seam of anthracite, or "metallurgic," coal targeted for midwestern steel mills.[6] The premise of *Coal TV* was that viewers would get an

Coal TV series: "Danger Runs Deep."

unadulterated look into the workings of the industry. A press release for the show touted the series as a way to "learn the truth" about a business as "American as apple pie," and its producer assured us that "authenticity is the key to this. We are not scripting these shows."[7]

The world we inhabit does not come unmediated, and *Coal TV* was, of course, no exception to that rule. Coal is not sexy stuff. The dullness of mining on a day-to-day basis clashes hard with the strictures of twenty-first-century media that demand, as Beers said, that the stories be made to "pop" in order to keep the viewer tuned in. The need for speed, tension, and titillation explains, at least in part, why docudramas like *Coal TV* have quickly replaced the traditional educational documentary on major television networks such as the Discovery Channel and the History Channel, those mainstream venues where history and reality are sold back to us as unscripted (or at least as scripted along factual lines). Docudrama is, however, an evolving genre with its own practices and norms, and it offers a type of "constructed reality" for us.[8] To produce *Coal TV*, for instance, required filming four hundred hours of a coal miner's life for each episode and boiling that tedium down to a short forty-two-minute sequence of (relatively interesting) conflicts, dangers, and obstacles. This type of extreme editing of mine work for popular consumption created a managed realism that extended to the selection of the location of the mine as well. Beers spent, he said, four years looking for the "right coal mine" to film—by which he meant an underground mine, rather than a surface mine, that was very small, privately owned, facing difficult odds, and open to letting his crew in.[9] Such parameters ruled out all of the big (environmentally devastating) open-pit mines, where the overwhelming majority of American coal comes from, and they ensured that the storyline could be cast as more personalized and melodramatic than otherwise. Similarly, the directors had decided that the average coal miner was not suitable for television either. As Beers explained to *Entertainment Weekly* and the hypermasculine media website *BroBible*, it took him "forever" to find the right men to film, coal miners who were "great characters" and who could "carry a story."[10]

Why *Coal TV* was produced in the first place was itself interesting. The motives were a curious mix of profiteering, hypermasculinity, and grotesque aesthetics. The series was part of a new reality show genre that centered on dangerous working-class jobs that were suited to what the *New York Times* called a type of "tough-guy TV"—a machismo television that "guzzles beer and pounds its hairy chest" and is "veined with blood and bleeps."[11] Beers is the uncontested progenitor of this new genre, and he was looking for a new subject. What made underground mining, in particular, exciting in this regard—and thus potentially attractive to the eighteen- to forty-nine-year-old male demographic that the series coveted—was the grotesque setting of the mine, which aligned with what the producer described as a horror-show aesthetic that drives media coverage of the everyday macabre, of such things as gun deaths, robberies, and natural disasters. As Beers explained, the coal mine was "schematically beautiful" in this respect. The lighting was easy to control, and the setting appeared to be, like a crime scene or the set of a horror show, "cold," "miserable," "dark," and "spooky."[12]

More than a million viewers tuned into *Coal TV* when it first aired in 2011, and as many as six million watched the series as it ran for ten episodes in the United States, before it was then rereleased in England under the UK Discovery Channel later that year.[13] That made *Coal TV* one of the more sustained exposures that a certain demographic in the United States (and their British counterparts) has had to the business of coal mining in the United States. *Coal TV* was thus a cultural event, in one sense—one that might have been unique in its gaudiness, if not its spectacle, but reflective (like reality television more broadly, which accounts for more than 50% of primetime programming) of some of twenty-first-century media's more generic tendencies toward the exploitations and erasures that prop up late capitalism and that fuel our affective affinities for fossil capitalism.[14]

The lessons of *Coal TV* become clearer if we look at how the series worked ideologically.

Fossil Capitalism as Object of Empathy

First, *Coal TV* made the ailing mineral economy—this hyped-up fossil economy that is reeling us toward planetary ruin—into an object of empathy. To a neutral observer, it might sound silly to grieve over a coal mine, to stress out over its failures, and to cheerlead for its success at a time when climate change from burning fuels like coal and petroleum is front-page news, but that is exactly what *Coal TV* asked us to do.

Coal TV was highly structured to identify with fossil capital from the start. The show closely aligned with the perspective of two major owners of the coal mine and thus gave us a look at the industry, as one press release put it, through their "eyes."[15]

Television, like the novel, permits a certain type of vicarious living.[16] It draws us into the backstories, the anxieties, the hopes, and the thoughts of its characters, and it has the added ability to position us quite literally behind the camera so that we can see the world from the perspective of a given character or a confidant standing near his or her side. Reality coal capitalized on this technology, using it to put us inside the heads of these mine owners, to allow us to pop in and out of their lives, and to let us merge affectively with their ways of thinking. At the beginning of each episode we are reintroduced to these two likable men, the CEO of Cobalt Mining, Mike Crowder, who we are told has "risked everything" to start up this mine, and his right-hand man, the firm's president, Tom Roberts, who tells us that he has always been "what you might call an entrepreneur." These petit capitalists stand in for capitalism writ large, and the coal industry becomes personalized through them, drawn down to a comprehensible scale that appears small, transparent, human, and humble. Absent is the bombast of the Koch brother's billions, the incomprehensible mechanisms of Halliburton's operations, and the Machiavellian politics of Massey Energy's behind-the-scenes operations in Appalachia. No mountaintops are being blown off here; no rivers are being offloaded with slag; and no political lobbying is going on as far as we know. The scale of fossil capital is perfectly human sized and seemingly responsible, and we are left to believe that anyone of us might start up a coal mine after we get tired of working at Starbucks or Target.

"I've been a millionaire basically a couple of times. . . . I wanted another shot at it," says the mine's president in the opening to the series. And that casual assertion that America is about "high" risk and "high" reward and that our class structure is perfectly fluid helps to align our aspirations with all of those wildcat drillers, small mine operators, and independent frackers who we imagine might be simple Joes like the rest of us trying their hand at panhandling in a land of opportunity. The casualness with which capitalism's deep structural inequality and risk is transformed into a sprightly plaything of free market agency—in which millions just come and go and wage workers sit side-by-side with owners as friends and near equals—permits all of us to knowingly nod with this CEO when he looks into the camera and remarks flippantly that "I feel like I'm going to make a lot of money, go broke, or end up in an early grave here." Perhaps our smiles are cut just a little bit short when the camera dramatizes that risk to his millions by showing us documentary footage of gas exploding out of a pipeline deep in the mine that sends miners scurrying for safety, as this careless stream of neoliberal signifiers equates the figurative death of an investment to the possibility of the literal death of less privileged miners.

Horkheimer and Adorno would have had a field day with this material. As early as the 1940s they could see how the media tended to situate the consumer as a passive subject and to identify upward, aspirationally, with management.[17] That was certainly the case with *Coal TV*. Here we voted, in a sense, for capital and for the boss; we stepped into the shoes and back offices of the mine owner and experienced the (staged) sigh of the investor at another loss, felt his emotional stress over having to mete out face-to-face disciplining, and looked and fretted over his accounting ledgers. Capital's losses are our losses, its sorrows ours, and its generosities a part of us.

This vicarious opportunity that *Coal TV* gave us to be with the boss—to understand the mine owner, the investor, and to empathize with him—was played out in each episode. One example will suffice. For instance, in episode 7, "Down 'n Out," we stand beside Crowder as he makes a difficult decision to fire a miner. We worry with him over the seriousness of the decision and understand his rationale. The nature of the firing is important in that it is not guided by the petty tyranny that otherwise hounds the working class and is not an instance of the union-busting layoffs typical of coal (including those that occurred at the Cobalt Mine shortly after the show ended) nor any of the brutal downsizing of the labor force because of market or stockholder pressures. This undisciplined miner, late to the job, shouting expletives, threatening people, and just plain intransigent, fits the trope of the irresponsible worker, and so we can agree with Crowder and Roberts that he has to go. We get, in this episode, a dramatic staging of capitalism's morality play (i.e., rational boss/irrational worker) that helps us to sit comfortably on the side of management: "You're fired," we hear, "because you fired yourself."[18]

More importantly, our identification with fossil capitalism, a commitment that is deeply rooted in our pensions and 401k accounts, for those who have them, and in a multiplicity of other mineral rites, gets played out in our need to make a profit on the mine and the anxiety we subsequently share with the owner about the mine failing financially. The main (if somewhat feeble) plotline of the series turns on the miners' trying to bring in seven cuts of coal a shift, or forty truckloads every twenty-four hours, to keep the mine "alive," so to speak. We are told week in and week out—in something like a theatrical performance of the conservative lament about a "war on coal"—that this mine is barely surviving and that it might fail if we don't make our targets. That drama of fossil capital gets thickly and crudely plastered for us throughout each episode:

— "It's do or die for Cobalt Coal."
— "They've got one month to turn a profit or they go bankrupt."
— "Tom and Mike are flat broke."
— "The next ten hours are critical for the mine."
— "Three and a half hours . . . lost equals nine grand, money that won't go to salaries or hooking up the generator."
— "Negative cash flow. The money is buried in the coal piles."
— "Seven cuts. That's a good day. $21K towards the 'genny' hookup and payroll!"
— "With Cobalt falling apart, Mike and Tom have to save the company."[19]

Profit is a moral imperative here, not something emptied out of meaning. Getting out of the red is tied not so much to the ethically hollowed-out rationale of returns on investment but to working-class jobs, working-class families, and the survival of a depressed mountain community. This lesson of capital's moral work is made explicit in the narrative that frames these episodes: "If Cobalt can keep it up," our narrator tells us in a deep authoritative voice in episode 8, "they might not only be providing jobs for these men but for the children of these men." Strangely enough, we find ourselves coaxing to life the uncanny dream that, if luck comes our way, we might have generations of coal to come.

So we root for the coal mine to make our seven cuts each shift and to get into the black for the team. This moral imperative to make a profit thus legitimizes the suffering we see. It rationalizes the risks that miners' bodies are subjected to. It justifies the stress they face from losing their jobs. It explains the drugs they take to ease their physical and psychological pain. And it authorizes the safety measures they cut to make targets. Our fetishizing of profit—this *objet petit a* after which we chase, this never enough and never achieved—in other words, helps to desensitize us to the human costs we witness along the way.

Of course, we have to take all of this on faith. The show only feigns to make us privy to the account ledgers for the Cobalt Mine. We hear about the account books in each new episode, and they appear as a regular stage prop in various scenes, but these proverbial books are also, of course, like all private capital, closed to us and perfectly opaque. The financing of the mine is flattened to this one imperative—seven cuts a shift. We, in truth, know nothing about the mine's fiscal situation: nothing about the wages of this non-unionized workforce, nothing about management salaries or bonuses, nothing about the money coming in from Spike TV, nothing about consulting fees, nothing about the investors who capitalized the company in Calgary, nothing about public relations and legal costs, nothing about what the owners have decided is an acceptable return on investment, and nothing about the capital tied up in technology. The so-called bottom line, these seven cuts per shift, is simply a natural fact rather than the social construction that we know it to be.

But let us play along. We are taught here that to make money and to create jobs means taking on the stress of the investor, it means getting in over our heads, staying up late at night thinking about personnel decisions, and making the capitalist's leap of faith into the future. This is the psychology, the self-rationalization, of the petit capitalist, and it informs us that this burden entitles us to, in the words of the series, "high rewards" for our pains. So we can cheer with the other owners when the men start making their seven cuts after so many nerve-wracking technological failures, growing debt, personnel problems, and environmental challenges, and we can join in on the huddle when Crowder looks out at the camera and tells us that it is finally our time "to be a winner, to make the touchdown, to score this time."

And for a while we do make our target. Good for us.

This empathy that is cultivated for fossil capitalism reaches its climax in the neoliberal trope of the corporate "team" (or, alternately, "family").[20] Coal TV lumps miners and investors alike together in what is portrayed as a family of trust and respect, where everyone plays their part and is celebrated for it. "We're trying to build a team out here," Crowder tells us in episode 7. "We're team players." This trope of the corporate "team"—a key neoliberal management strategy that doles out performance plaques and bromides of inclusion rather than financial security—provides the emotional substructure to Coal TV, and it works to transform the ailing mining economy (this vast complex impersonal and devastating thing) into a sentimental object. That trope gains emotional substance in the way it is casually merged with the show's discourse of family, which is showcased in discussions of business, responsibility, trust, and risk. Perhaps most egregious in this regard is *Coal TV*'s introductory sequence, which borrows its cultural capital both from the literal family of sons and brothers in the Appalachian mines and from the brotherhood of a working class that looks out for each other, thus appealing to the solidarity of blood and the solidarity of oppression to authorize the corporate family. In that sequence, a montage of working miners' faces streams across the screen to feverish violin music, accompanied by the narrator's punctuated phrase, "Sons . . . brothers . . . fathers," and then after some pause, "COALMINERS!" This remaking of the family in the coal mines permits us to imagine that the miner and the president, the foreman and the boss, all share a certain power and affective affinity with one another in this enterprise, minimizing the direction of power, its raw execution, and stripping it of its social locus.

There is a quaintness to it all. Fossil capital appears not as a mechanism of social privilege and stratification, not a complex business of differentiated power tied up in the modern world's dialectic of misery and pleasure, where some lose millions and others count pennies, but rather as this homey business that operates locally, humbly, sentimentally, and without violence or prejudice. Such a portrait makes it easier for us to cheer for our retirement accounts, to cheer when the industry has rebounded in the next financial quarter, without having to pay much attention to the train of disasters in its wake.

The Coal Miner as Object of Consumption

If *Coal TV* made fossil capital into an object of sentimental attachment, it also invited us, without making us get up from the couch, to exploit the body of the coal miner—to consume that body not only in the old material ways, as we do when we turn on the lights or get into eighteen hundred pounds of automotive steel, but in a second way that doubled down on that exploitation by transforming the precarity of working-class life in the underground fossil economy, the reality of men's bodies getting used up in the mines, into a public spectacle for entertainment. The exploitation of physical labor has always been structured into the fossil economy, and here it was made visible, legitimized, and prescribed for us to swallow.

Coal TV was something of a modern-day peep show into the hard lives of Appalachian coal miners. Although a small subset of the show's audience came from Appalachia (as is evident by the intriguing debate in the blogosphere among local miners over the authenticity and the exploitativeness of the show), the vast majority of its audience of six million viewers did not. *Coal TV* invited that majority, who were not "of coal," thus to participate in a type of class and regional voyeurism that entailed watching other men wake up and slog to work in the morning, stress out over making mistakes and possibly losing their jobs, and suffer from occupational health hazards, such as dying quickly or slowly on the job from a mine collapse or respiratory, kidney, and heart diseases—all the while with chips and beer in hand. The structure of play here was, in other words, tantamount to screening footage of East LA sweatshop workers for the pleasure of Lord and Taylor shoppers on the West Side.

As one West Virginia blogger put it: "There is something weird about *Coal* the TV show."[21] That might have been the understatement of the season.

Context is important. Coal is not casual in McDowell County, West Virginia, where *Coal TV* was filmed. It is embedded culturally in community memory, working-class resilience, and a regional melancholy carried in bodies and minds from generation to generation, and it is anchored in a political and economic structure that discourages any real incentives to transitioning away from coal. This is where JFK campaigned in 1960 to take up the war on poverty among white workers, and it is the place that LBJ had in mind when he began to wage that war at the federal level.[22] The state's coal lobby still has undue political power here, and it bullies the state with the message that "a vote for the industry is a vote for yourself, your identity, your survival," even while its big players, such as Massey Energy, have shown a callous disregard for workers' safety and the region's long-term economic prospects.[23] The coal industry has, in other words, not been kind to McDowell, not been an engine of democracy, and not been a means to sustained upward mobility or to community health. In the twenty-first century, McDowell's poverty rate still hovers around a staggering 36%, nearly one-quarter of the area's population is disabled, 35% of the county's adults lack a high school education, and life expectancy for men stands at sixty-seven, a chilling ten years less than the national average.[24] Moreover, the Cobalt Mine was itself a union-busting mine, giving to the workers here a special degree of contingency.[25] If fossil capital promises mobility in theory, it has a century of failure to account for in West Virginia.

Filming the mines in McDowell for entertainment is thus a little complicated.

Coal TV both acknowledged that context and exploited it. On the one hand, *Coal TV* empathized strongly with the Appalachian coal miner, and thus it held out the possibility of closing the empathy gap between the producer and consumer of underground coal. The miners in the series were portrayed as something like an extension of ourselves, as the best of laboring people: sincere, hard working, and deserving. And the series offered us, to some extent, a chance to see the world through their eyes. We hear, for instance, their perspective in snippets played here and there throughout each of the ten episodes. Miners tell us that they go into the mines because they lack options: as one man puts it in the opening sequence, "They're ain't no jobs in West Virginia. It's either diggin' coal or flippin' burgers, and I ain't no burger man." They tell us that they tolerate the danger and the reduced life expectancy coal mining brings with it because their families need them to: "I'm down here," one miner in episode 3 says, "for my family. . . . I love 'em a lot. I love 'em enough to come in here for 'em." Similarly, they tell us that some miners take pain killers like Oxycodone to deal with pain from injuries on the job, such as one man who in episode 3 explains that he got addicted to the drug after hurting his back in the mine. And they explain that, like every one of us, they are just doing their best to get somewhere in life: "We're just country people," the miner Jerry Edwards says in episode 2, "working to give our families what we never had, and try to make better for our families."

This empathy for the coal miner is played out in the way that *Coal TV* zeroes in on one skilled and kind veteran miner, the late Andrew Christian Sr., who is set alongside the CEO, the president, and the boss (i.e., foreman) in the opening sequence to the series. Christian is afforded a certain celebrity status in the series, given the nickname "the Legend," displayed in bold letters across the screen at the start to each episode to frenetic country music (as if calling out a star for an NBA playoff game), and placed on something of a par with the owners of the mine. We are told that he has earned his legendary status and stature for his skill in making difficult cuts in barely lit spaces less than forty-two inches high, and we are informed in several episodes that he has once again "saved" the mine from going under when lesser men fail to make the targets set out for them by management. *Coal TV* makes an effort to humanize him, to show him to be more than simply a worker. We learn a little about his family life, for instance, in episode 4, when his son, Andrew Christian Jr., crouching beside him in the mine as an apprentice to his dad, worries about his dad's health (because the father is suffering from kidney disease) and imagines that he might one day take his place: "I am not the best," he says humbly. "But I will be." *Coal TV*, in other words, opens up the possibility that getting to know these miners—not as consumable objects but as people—might cause us to wonder if it is morally ok to tolerate their exploitation, to accept the abuse to their physical health that coal mining inflicts, and to accept their financial hardships for the sake of a mass market in cheap steel and cheap electricity.

This is the best of *Coal TV.*

But such episodes of enlightenment—these humanizing moments when we clearly see the stress and injury of the fossil economy and when we consequently begin to sweat a little over our culpability—are quickly shuttled aside so that we can enjoy our tour of the mines and remain anesthetized to this dialectic of privilege and pain that defines life under the terms of the fossil economy.[26]

Empathy, of course, requires identification and humanization, and objectifying other people tends to work against its solicitation. *Coal TV*, for its part, worked to objectify (as well as commercialize) the coal miner's body in several ways.

Most egregiously *Coal TV* indulged in a "pornography of class violence" that turned the pain and risk of these Appalachian coal miners into spectacle for entertainment without providing any meaningful ethical or political context that could have redeemed it. Of course, capitalizing on people's suffering—on gun deaths, on robberies, on fires, on crying mothers—is the stock and trade of the twenty-four-hours news media under late capitalism, and it has become an accepted part of a global consumer society, which takes intimate aspects of people's lives, including their private joys, their grieving, and their interpersonal relationships, and serves them up for public consumption. So *Coal TV* was representative of our culture rather than unique in this respect. Moreover, we might argue that the public display of violence—that is, making visible an injury that was previously hidden from us—can be done with a purpose. Karen Halttunen has shown, for instance, that white northern abolitionists indulged in a certain humanitarian pornography by displaying the flogged body of slaves in their propaganda for political and moral objectives.[27] But this pornography of pain can also be purely exploitative, as real human experiences are flattened to simulacra—the play of symbols for consumption—and detached from reality or a politics that might help us understand the human dimension of those experiences.

Coal TV turned the history of death and disability in McDowell County into simulacra—into a plaything.

The main tease of the show (the word is obscene in this context) was the death of a coal miner. Each episode held out, in its own way, the possibility that the viewer might get to go down into the mines to see a miner die in some spectacular fashion. "I've seen people git keeled," one miner, nick-named "the Wildman," says in thick Appalachian dialect in the opening credits to each episode, and that one-liner, along with endless other iterations on the theme of death and injury, was recycled ad nauseam throughout the series. The producer of the show claims that he had, in fact, intentionally targeted the "most dangerous jobs" he could find, using the list of industry accidents and mortality provided by OSHA (Occupational Safety and Hazard Administration) to identify them, and thus this project was in keeping with his more flagrantly lurid shows like *1000 Ways to Die*, which titillated viewers with stories of unusually macabre deaths. And the series capitalized on stories of death and escape, such as the contemporaneous entrapment of miners in Chile and a nearby mine disaster in West Virginia. Filming the health of and the safety risks faced by coal miners under the guise of reality documentary was simply an innovative way of capitalizing on working-class morbidity.[28] But, to be sure, it was death—the shortened life span of the Appalachian miner—that was being sold as theater.

Take the following sequence from the series' trailer:

A NEW SERIES THAT WILL TAKE YOU TO NEW DEPTHS OF DANGER [Dark ambient music.]

"I've seen people with their arms broke, their legs broke, their backs broke." [Miner talking.]

"I've seen people die. I've seen rocks fall on people." [Slate falling next to two miners' heads.]

"If you think your job's tough, come work in the coal mines." [Miner talking.]

"Rocks can crush you." [Rocks cascading.]

"Machines can kill you." [A large machine coming at a miner.]

"And gases can choke you to death." [Miners yelling "Oh, shit!" and running from a gas explosion.]

"But it's kind of an adrenaline rush, and I'll never stop doing it." [Triumphal music begins.]

"I love living on the edge." [Young miner talking.]

This theater of death and disability (and its rationalizations, as we will see) was performed in each episode of *Coal TV*—even as the tease always remained out of reach. Gravestones with the word "FATHER" transposed over miners' accounts of men killed, the door of an ambulance closing, falling slate with miners scurrying: *Coal TV* made one community's experience with premature death and disability into an object of entertainment to be dangled in front of viewers, something to chase after, in order to keep us tuned in.

We have not traveled so far from the public hanging.

Consuming death as spectacle requires, however, that we be authorized to enjoy it, that human loss be diminished and rationalized for us. We need to be freed, in other words, from death and disability's political context; people need to be objects. *Coal TV* gave us permission to enjoy working-class suffering, and it did so by naturalizing the social order of fossil capitalism and by anesthetizing us to the inconvenient fact that today's dialectic of privilege and pain is rooted in class (and regional) inequality.

The most obvious way it did this was by marking the bodies of these workers as being different from those of viewers at home, as being *other*. *Coal TV* traded on the dirt associated with coal to present the miners as being stained, as alien, or as foreign. White teeth gleaming against coal black

The coal miner's death as hyperreal spectacle.

lips centered on frame. Seams of coal engraved in wrinkles on a miner's body in a close-up. A rough hand caked in coal reaching out from a body off screen. Eyes, lips, hands, arms, necks—pieces of these workers' bodies were isolated for us and aestheticized for their roughness, dirt, and poverty. The miners in *Coal TV* appear, that is, and this is true even when they are being interviewed in settings outside of the mine, as having coal dust grooved into their bodies—as being sooty by descent, the stain of coal establishing a certain distance between the viewer and the viewed.

Reinforcing that distance was the producers' use of Appalachian dialect as a second marker of difference. Dialect is something we all have, of course, and it can evoke a strange admixture of attraction and repulsion. *Coal TV* capitalized on the strangeness of Appalachian dialectic by all too often whittling down the various miners' interviews to a few short ungrammatical sentences in thick dialect. In effect, it differentiated and truncated the voice of these miners by turning their speech into consumable sound bites set in a semiforeign tongue. The clearest example occurs in the conclusion to the show's introduction where a coal miner being interviewed, his face coated in coal, has his entire life philosophy and politics reduced to six words: "'Bout all I know minin' coal."

And that is that.

Objectifying class and regional difference.

Marked as different in body and speech—tied to the ungrammatical, the vernacular, and the quaintly folksy—the coal miner verges on being *thingi-fied*. But perhaps as important is how *Coal TV* chose to use miners' own words, their own self-rationalizations, to legitimize the inequitable social order in which McDowell men find themselves—this corner of fossil capitalism that objectifies workers' bodies economically. If we had any qualms about watching other laboring people with limited life opportunities perform dangerous work with the expectation of reduced life expectancy, then *Coal TV* let us know that we need not worry about that. Coal miners, we hear, love this stuff. The short quotations that the show recycles most frequently (as exemplified in the trailer) come from one or two young miners who tell us that they "love living on the edge" and that they get "an adrenaline rush" from the work. That message is reinforced by a second one: that Appalachian coal miners were born to mine, that they belong in the mines. "This is where I belong," one miner says. "It's just in my blood," says another. And the last words recycled in the series' trailer are from one young man who says: "If I die, I hope they put me back in the ground *where I belong*" (emphasis added). The bravado and the sense of belonging expressed by miners is, of course, not so much false as it is representative of a small part of a whole. The sin here is one of omission (or emphasis), not untruthfulness. For every young miner who gets excited to go into the mines there is an old miner with emphysema and a mother who regrets that mining is about his only option. The effect of such editing is thus political. By recycling certain representations and not others, the show makes the job of underground coal mining sound a lot like playing football or hang gliding in one's free time. Some of us get excited about risk, and some of us were born to do this stuff.

But if we don't believe the men, we also have the assurance of management. Management too tells us that capital's inequality and its risk are natural and that the global economy just works this way, and that that can't be helped: "People do get hurt," Roberts acknowledges. "And you have to accept that."

The impression that is created by this type of editing of interviews is that working-class risk and inequality are part of the natural order. Men don't have to do this type of work; rather, they are free agents who choose to do so, and the risk they face is just a fact rather than a social construction. The human costs of fossil capitalism, in other words, are whisked away, and we are permitted to regret that some bodies are put on the line for the sake of progress while being assured that there is a natural order to the fossil economy that simply isn't within our power to change. This objectification of other people is one of the persistent human failures across history, and it has reached disquieting depths in the neoliberal economy of the twenty-first century.

What is not said is that miners at Cobalt worked in the face of constant threats to life and limb in an industry that does not frequently abide by its safety regulations, that renders workers contingent, leaving them without job protection or security, and that does not, of course, provide workers equal (or even reasonably proportionate) rewards for the risks they take to maintain the fossil economy. What objectification means in McDowell County is that people hang on to keep the community together, flipping burgers and digging coal. But, to be clear, there is nothing natural about this. The occupational constraints, the contingency, and the difficult conditions in McDowell are a modern product. They are the product of the conscious and unconscious reproductive choices of our capital investment in a mineral economy that we collectively create through our mineral rites.

Coal TV presented the consumer with a potentially transformative collision between two classes of people in the fossil economy, the producer and consumer of coal, but it then invited us to ignore that relationship. The consequence was that the body of the coal miner was doubly capitalized on— reified twice over: first, in its use value in producing the cheap energy we need for our television screens, and second, in its entertainment value. In other words, the plight of exploited labor becomes under the terms of reality TV a distant and aestheticized thing—something we can enjoy for a bit and also feel a little bad about before we move on to *Lobster Wars*.

Fossil Capital and the Erasure of Politics

Third, and finally, *Coal TV* helps to exemplify how mainstream culture naturalizes the fossil economy through countless small and large editorial decisions that create certain silences and erasures that make its social and ecological costs appear a little more palatable to us. That is, we can see negotiated in this case, as in every media decision about how to represent labor, trade, play, and production, the "tension between the knowledge and ignorance," as Arjun Appadurai puts it, that goes into shaping our understanding of the commodities in our lives.[29] This conscious crafting of knowledge and ignorance in the case of the media, or what we might think of the production of the visibility and invisibility of commodity production, is political. Entertainment might under private capital pretend to be agnostic politically in its representation of the world we live in—offering to suspend us in a contextless present where ideology and politics disappear on the screen to accommodate viewers of all persuasions—but such things never really do disappear. And *Coal TV* was no different in this regard.

The first lesson in this respect concerns how knowledge about coal is distributed today, and, in particular, how the interplay of public knowledge and public ignorance shapes how we imagine the contemporary landscape of the fossil economy.[30] To *see* our dependencies properly—to mediate ethically between the sites of energy production, those of consumption, and the spaces in between—means being able to paint a meaningful and representative portrait of the mineral economy that will enable us to act, to visualize and contextualize it properly. The difficulty is that the opaqueness of global capitalism (and fossil capital more specifically)—that is, the spatial, social, legal, and epistemological distances that divide consumers from the various links in the chain of production and exchange—makes it difficult to comprehend the impact that our material practices have on peoples and ecologies along the way. But to start, we might begin with the point of production: the coal mine, in this case.

Seeing the coal mine properly presents a challenge. Not only do the account books and internal correspondence of the coal industry sit behind locked doors but the actual sites of production, including the industry's massive slag pits, its strip mines, its chemical plants, and its power plants, are frequently off limits to the public or sited in remote locations that make our mineral practices less likely to be witnessed by most consumers. Sometimes the opacity of carbon production is by design. The Austrian artist Ernst Logar learned this, for instance, when he spent two years seeking permission to photograph some of the big oil operations in Europe and ended up, for his troubles, with a long string of rejection letters thanking him kindly for his interest. Logar had to accept that his perspective on the oil field would come from behind the fence. Likewise, the producer of *Coal TV* explained that he'd "learned over the years that you can't get into a big corporation," so, in his case, had to settle on a story of coal mining that centered on the small mine.[31] Other times that opacity (and the ignorance it reproduces) is a by-product of the social and spatial distance characteristic of globalization. For example, even those industries that open their operations to the public, such as Fort McMurray, which offers guided tours of oil drilling and mining in Alberta's tar sands, are still inaccessible to most of us, because of the long distance traveled by the commodity and the resulting spatial separation of the consumer and producer. For the handful of visitors who do make it to the pit, a second hurdle presents itself—the industry's daunting scale of extraction and the temporal nature of the changes under way make it virtually impossible to capture effectively in frame, whether by photograph or other artistic rendering.

To fully understand it one has to be there. And most of us are not.

That made *Coal TV* potentially transformative. The cards were lined up nicely. They included a team of skilled cameramen with several months of wide access to an operating coal mine, cooperative mine owners, and the rights to broadcast for ten weeks on a major television network to a paid subscription base of nearly a hundred million households in the US and Canada. *Coal TV* would give us, in other words, as its producers promised, "a never before seen look at the profession of coal mining."[32] That gave to the series the real possibility that it might redistribute knowledge about coal in a catalytic way.

But it didn't really do that, or it did that in a very proscribed manner. We were led down a rabbit hole.

Not much new appeared on *Coal TV*. This was because editorial choices failed to challenge our received image of coal. The main reason the show failed in this respect is that the shaft we traveled down was not representative. It was something like a remnant mine, a historical outlier that did not reflect the dramatic changes that coal has undergone in the last seventy years. Today, most mines, in contrast to the Cobalt Mine, do not operate underground in the United States, nor do they operate on this puny scale. Coal today is gargantuan, impersonal, and climate altering. It is not humble, not homey, and not familiar. Even apologists for the industry took issue with the series on this score. The proindustry magazine *Coal Age* complained that "if the series was actually going to do the U.S. coal industry justice, it would have rode in the jump seat of a 400-ton haul truck in the Powder River Basin with a female driver. Or, for the best underground experience, it would have followed longwall miners at one of the 50 faces in the U.S. where it could have filmed coal being cut safely at a rate of more than 5 million tons per year." The point is that Cobalt's forty men and seven cuts a shift give us a misleading idea of the size and impact of the industry. If we see the Cobalt

Mine as representative of corporate coal, as the series encouraged us to do, we are deceived into thinking that it is an industry that is an awful lot like the industry of our grandparents, one that looks, dare we say, sustainable.[33]

Coal today is, of course, not the coal of our grandparents. Coal's alarming scale stands in stark contrast to the miniaturized work done at the Cobalt Mine. The biggest mines, like Peabody's North Antelope Rochelle Mine in Wyoming or even Consol's more modest underground Bailey Mine in Pennsylvania, extract between twelve and ninety-three *million tons* a year. Cobalt, in contrast, set as its target a mere eighty-four thousand tons a year, a tiny output that amounts to less than 1% of these major players. None of the top fifty coal mines in the United States, in fact, produces anything less than four million tons a year.[34] Twenty-first-century coal is, in other words, of theatrical Greek proportions—big, Bacchian, and bombastic.

One reason that the scale of coal today is so different from that of the coal of our grandparents is that two-thirds of it is extracted through surface mining. Although most of the big surface mines are located in the American West, concentrated in eastern Wyoming, a third of the mining in Appalachia is a version of surface mining called mountaintop removal that is especially thorny both politically and ethically. Today, miners in Appalachia don't simply disappear into a local mountain for an eight- to ten-hour shift and come back lugging sixteen tons of coal to the surface, to quote the late Tennessee Ernie Ford. Instead they are paid to do the much safer, and environmentally catastrophic, work of removing the mountain itself. This type of geological upheaval defies the limits set by the older muscular economy in the mines. Coal mining is now so thoroughly mechanized around big machines, in fact, that the entire industry only employs about 52,000 men and women in the United States, or about 7% of the 720,000 men who once mined coal in the 1910s. Modern-day coal mining does not, that is, provide many jobs (even if it provides crucial ones in places like McDowell), and it doesn't leave a small hole in the ground.[35]

If we keep our sights trained on Appalachia, today's surface mining practices have resulted in the removal of five hundred mountaintops since the 1970s. The parts of that biologically rich mountain range that have been extracted are now 40% flatter and ecologically restructured so that they only support grass and not forest.[36] What industry and regulators call overburden (i.e., the part of the mountain that sits on top of the coal) has to be removed in surface mining before extraction, and thus what the rest of us call earth—soil, mosses, grasses, shrubs, trees, mugworts, lichen, microbiota, and the animals of a region—gets upended and then dumped, broken, and jumbled, into nearby valleys and streams, only to be reassembled later into what regulators call the earth's approximate original contour. Flying southern squirrels, old-growth pine, and mountain memories are gone in one fell swoop, or, to be more precise, in one gargantuan scoop after another. Appalachian coal mining amounts, in other words—and this is true despite recent federal efforts on behalf of the Obama administration to limit surface mining here—to the systematic dismantling of one of the nation's oldest ecosystems step by step, light switch by light switch. To know contemporary coal one need only look to the aftermath of the three-hundred-million-gallon coal slurry spill in the town of Inez, Kentucky, a stone's throw away from the Cobalt Mine, that drowned a town and its rivers in toxic sludge.

Coal mining today is thus an existential issue—tied up in weighty ethical questions about what it means to remove ecosystems permanently and to alter the geological contours of the planet.

But seeing coal correctly is not just a matter of appreciating the acute threat to immediate environments and the people who rely on and identify with them. It also means, of course, acknowledging the main, and defiant, issue of our lifetime—global warming. Because coal accounts for a quarter of greenhouse gas emissions both in the United States and the world at large, coal mining is now inextricably bound up in moral questions about what type of life we will leave to our children. Knowing coal thus means knowing this context, and that is not something that *Coal TV* decided to provide. Instead, *Coal TV* swapped out the real debate over climate change for a false dilemma. According to Spike TV, the only real "downside" of coal mining is that "like most fossil fuels, it's not easy to get to, plus transportation of the fuel is astoundingly difficult."[37] It is as though, for this network's producers, the challenge facing humanity is figuring out how to mine and transport coal economically. Arguably, it takes a certain effort to film four thousand hours of documentary footage of coal mining in the twenty-first century and not touch on the issue of global warming.

Climate change is not a marginal subject in rural energy communities. Despite popular perception, debate over global warming is not the province of a liberal urban intelligentsia but a fundamental part of local discourse in most energy communities across the United States. Polls show that 69% of Appalachians, in fact, contend to know a great deal, or at least a moderate amount, about climate change, and that as many as 42% of Appalachians believe that climate change is already on us and that it is the result of human practices.[38] Moreover, driving down the turnpike in the eastern coal regions from Pennsylvania through Kentucky one would have to work hard to ignore the parade of propaganda, from both the left and right, over coal and climate change. Highways are peppered with billboards that, on the one hand, tout the promise of wind power as an alternative to coal in the region, and, on the other, downplay the reality of climate change and coal's contribution to it. For a couple of years, for instance, one could not drive through Pennsylvania's Allegheny mountains, near its coal beds, without confronting a giant clown's head, with a big red nose, smiling idiotically, to the message "He believes in global warming. Do you?" Presumably the snow on the ground in these hills was proof enough that we were in the midst of a great hoax. Thus it takes work to ignore the issue.

Being apolitical—for the sake of agnosticism—produces its own ignorance.

Coal, in other words, presents us with a representational dilemma. When we choose to see coal as a small underground operation like Cobalt's, we add fuel to the status quo of the fossil economy; we minimize the human and environmental costs of today's practices and provide ourselves with an escape hatch that allows us to avoid thinking about what it means to plug in our iPhones or replace our next car with another eighteen hundred pounds of steel. Perhaps things would be different if the image we had of coal was of rivers being buried, hollows being filled in, species being lost, communities being abandoned, and our grandchildren slipping downward on modernity's social scale as the world heats up.

Or maybe not. But at least we would know.

The final lesson to take from *Coal TV* concerns the increasingly important discourse of risk under fossil capitalism. Risk is an inevitable part of the world we live in, and every culture accepts different types of risk as part of its worldview. But what risk means to people and how its complement, catastrophe, manifests concretely in any given time and place is the product of social structures and human decisions, and under fossil capitalism, risk (and its complement, the likelihood of catastrophe)—whether in McDowell County or post-Katrina New Orleans—manifests quite predictably, if not neatly, along class, racial, spatial, and gender lines. Risk has, in other words, a distinct social composition. Mitigating this risk and its stratification presumes that we can learn how to see, assess, define, and represent that risk effectively. What makes that complicated, however, is that the power to predict risk and direct its management rests firmly in the hands of the privileged. As the risk theorist Ulrich Beck tells us, some actors have a "greater capacity to define risk than others."[39]

Risk is an unwieldy signifier, and its referents get lost in this discourse. For instance, there is the risk that the venture capitalist (or even the small-time 401k investor) undertakes in putting his or her money in a company and banking on its future. This is the privileged category of risk that Cobalt's Mike Crowder, Tom Roberts, and its various unseen shareholders undertake. The complement to that is the risk that the working class undertakes when it puts its bodies, rather than its money, on the line. At the Cobalt Mine, this risk takes the shape of the reduced life expectancy that Appalachian coal miners accept as their condition of employment. Beyond this, the permutations of risk expand outward structurally to a larger circle of stakeholders—including neighborhoods, towns, counties, states, countries—that accept collective risk when they invite, or fail to prevent, large-scale extractive and heavy manufacturing industries from taking up residence nearby. Such risk has already been brought to light in relationship to industrial sacrifice zones in Kaohsiung City, Hebei Province, the Navajo Reservation, and Inez, Kentucky. And, finally, there is, for our purposes, what we term *climate risk*, that is, the forward-looking calculations that insurance companies, hedge fund managers, manufacturers, states, and world governance bodies make when they try to anticipate the costs and opportunities that modern institutions and populations face from the unfolding ecological crisis. None of these categories of risk is identical to one another. Each, however, plays a part in modernity's "risk society," where what we call living is increasingly framed through the prognostication, anticipation, and quantification of risk probabilities in the future and by the inevitable catastrophes and slow violence that will result from defining and acting on risk in certain ways and not in others.

Arguably, neoliberal capital has had the upper hand in defining risk in the twenty-first century. It thus has had a privileged capacity to mitigate that risk to the extent that it is foreseeable. Risk in a neoliberal framework, however, amounts to the art of managing financial opportunities and protecting investments from factors that might impact it, such as the probability of disruptions from labor discontent, natural obstacles, political instability, underdeveloped markets, climate change, and so on. Neoliberalism thus repositions most human and ecological costs of fossil capitalism as *externalities*, redefining them as factors supplemental or tangential to rather than constituent of the short- and medium-term calculus of its risks and rewards. It makes them moral accessories.

We too, as minor players, live inside this culture of risk. For instance, we embrace its ethos each time we designate our retirement funds as being aggressive, moderately aggressive, or conservative. We participate in its drama each time we read through the *Wall Street Journal*'s *Morning Risk Report* and look at its global risk map. And we accept its emotional ecosystem each time we write off today's global structures of economic inequality and precarity as unfortunate but unavoidable. It is difficult, even when we try, to escape this way of living inside capital. Our newspapers tell us that "all capitalism is about risk" and that "rewards do not come otherwise."[40] Our university textbooks socialize us into this culture of risk by teaching us that the "tradeoff between risk and return" is one of the ten principles of capitalism.[41] And our global consulting bodies advise us that "calculated risk taking is a fundamental precept of capitalism: no risk taking, no innovation, no competitive advantage, no shareholder value." It is axiomatic enough that the International Risk Management Institute puts it crudely this way: "Taking Risks to Create Value—It's What Capitalism's All About."[42]

Of course, if one operates from inside a neoliberal discourse of risk, one reproduces the worldview of the 1% and the worldview of those of us invested enough in, or deceived enough by, that 1% to hitch our fortunes to it. When we accept its premises, we also accept that our relationship to the future can be modeled through a calculation of projected returns and losses and that external costs, such as ecological disaster and human precarity, can be addressed, or will be addressed in the future, adequately through this logic. Looking at how *Coal TV* engaged this trope of risk can be instructive; it shows how fossil capitalism as it is practiced today fails to account for risk as an incommensurable and moral object—and thus how it undermines our ability to act properly toward the future.

Coal TV framed coal mining in Appalachia through the axiom of risk and reward. The starting point for each episode was that in this industry "the risks are high . . . but so too are the rewards!" That trope was recycled throughout the series, and it grounded the emotional habitus of the show. In episode 1, "Master of Mines," for example, the stage was set for the entire series: "Two weeks ago," the narrator tells us, "Tom Roberts and Mike Crowder, a former computer CEO, took over the rights to mine this site. For both men it's the *gamble of their lives.*" Crowder testifies to that gamble: "We are really *rolling the dice,*" he says, "and we're *gonna sink or swim* here in the next thirty days." This language of capitalism's risk is showcased and recycled in myriad ways at the start of each episode, and it informs the overarching drama of the show. It is given added visual heft by the camera's positioning of us over Crowder's shoulder so that we can examine an Excel spreadsheet with its debit column listing out the long string of debts that the owners are incurring from running the mine. "We're spending a lot of money right now," the CEO tells us, "more money than we have. That's why we're counting on the cash flow to fix our problems here." This lesson in indoctrination ends with the narrator clarifying that the risk to Crowder and Roberts is very large—another $4 million for mining equipment and for hiring thirty-five local miners, we are told. They are spending everything they've got, and they are thus all in: "It's a new ballgame now."

This is capitalism's drama: investors risk their money, they have sleepless nights, they create jobs, and so they deserve high (in fact, astronomically high) rewards in comparison to workers. But risk in this ideological sense, as capitalism's burden, its play, and its just desserts, absorbs into its logic other existentially distinct types of risk. In particular, it reclassifies the risk to workers' bodies, to local ecosystems, and to all of us from global warming—these morally incompatible things—as part of an accounting logic that erases the exigency of their circumstances.

Take the following sequence from the series' introduction, which repositions working-class occupational risk within this neoliberal discourse of high risk and high reward. A voiceover begins by telling us that "the risks are high" in coal mining. This is the drama that the series builds around the investors' financial gamble. But instead of being titillated by dollars plunging down the drain we are titillated by a dizzying assemblage of close calls to working men's bodies down in the mine itself. These class risks are piled up quickly on top of each other to music in frenetic tempo: a rock ledge falls next to two coal miners sending them scurrying to safety; a gravestone appears ominously beside a miner; a miner's eyes dart worriedly, aware that something has gone wrong; another miner sees an accident coming and yells "Oh, whoa"; the mine's foreman looks out at the screen and tells us that he has seen people get killed in the mines. The sequence ends with the door of an ambulance closing, letting us know that injury and death in the mines are real experiences for the working class. In this fast-paced montage, the discourse of high risks to coal miner's bodies gets conflated, however, with the risk to investors' money.

The neoliberal logic at work here, which absorbs working-class injury into capitalism's raison d'être, finds its conclusion in the premise that we can expect high rewards for taking these risks. Here we shift abruptly from the imagery of miners' bodies at risk to Cobalt's president, Tom Roberts, looking up at a huge mountain of coal, smiling and laughing to himself, in recognition of the high rewards that such risk earns us. That joy is doubly reinforced by a worker climbing onto a different pile of coal, raising his arms in victory, and yelling "Yeah!," as though he too is celebrating the high rewards of coal mining. (This latter image was repurposed from a scene where a worker was celebrating his defiance of management, not celebrating management bonuses.)

Capital's logic of risk functions in this framing sequence, in other words, to answer to unrelated referents and to make a value claim that is logically incongruent. This is how the discursive logic of risk works under capital. We have in miniature fossil capitalism's rationalization of its dialectic of injury and health being played out in a logic that conflates the mine owners' financial risks with the miners' corporeal risks and flattens them both into one accounting, into a single metanarrative. We are taught, in other words, as we are taught in life under capital more generally, that losing one's investment is not so different from losing one's life and that a symbolic mountain of coal—this promise of high rewards, this metonym for the dollar—represents adequately the pain of the worker and investor alike. But if this is our moral universe under neoliberal capital, then something is not quite right.

What begs asking are the questions that a neoliberal discourse of risk never allows us to answer. Is this mountain of coal worth the reduced life expectancy and the broken backs and knees of the coal miner? Does risking one's money earn one the right to significantly higher rewards than those who have only their bodies to risk? Does risk management, under neoliberal accounting, adequately address the moral obligations we have to each other and to the future? And can fossil capitalism address climate change?

We know the answers to those questions—and so we can't ask them.

To bring this essay to a close: there was nothing really wrong with *Coal TV*. Not at all. Not even a little bit.

The logic of *Coal TV* is the logic of fossil capital under neoliberal management. It is the logic of a spastic consumer culture that dares to commoditize even the pain and disability of the world's working class, and it is the logic of an amoral and immoral capitalist order writ large, albeit played out on a small stage with a small cast of characters. *Coal TV* is who we are under neoliberalism, and it clarifies who we cheer for under fossil capitalism. It speaks for a politicized people without knowledge of politics, a historical people without knowledge of history, and a future-oriented people with no plan for the future. It counsels us, like it counsels the underground coal miner, to keep our nose to the grindstone and to simply hope, since we know nothing else, that we can pull our seven cuts and keep the mine alive for one more day.

Six

Carbon Culture

How to Read a Novel in
Light of Climate Change

To read a novel in the midst of a global crisis poses certain problems.

Time is short, and climate change begs for immediate action. So, with good sense, we turn to the market and to the laboratory for answers, giving to modernity's privileged knowledge brokers—the businessman, the engineer, and the scientist—the first shot at defining and solving the problem. So much makes sense.

But the peg moves forward and the anomalies pile up. Today's action on climate change comes haltingly, and it is conducted inside of prefabricated boxes. For each step forward, for each small breakthrough, we experience a dizzying array of distortions, head fakes, and reversals. The problem is that climate change is not primarily a crisis of pricing or engineering, although it is also that. Climate change is a crisis of will, a crisis of values, empathy, and of intellect—in short, a crisis of imagination.[1] We are rutted in, what we might call, *cultural path dependencies* as surely as we are in moribund energy networks.

So how do we overcome this fossilized inertia?

Reading fiction and writing fiction have their part to play. Climate change demands new subjectivities, and tools like these will be needed if we are to exercise creaky joints that have not moved for years—if we are to walk, think, and breathe in more sustainable ways.

Let us take literature as an example. Reading fiction is not a Sunday morning spent among oranges, coffee, and a cockatoo—a passive thing of leisure. It is a cultural technology specific to our species, honed over at least five thousand years, that allows us to communicate human emotions, dreams, and values across time and space. According to evolutionary biologists, reading and writing are a later development in the larger cognitive revolution that gave *Homo sapiens* a competitive edge over other humans like Neanderthals and Denisovans. By developing expanded networking capacities, we gained the ability to join our memories, beliefs, and dreams to one another. That made it possible to act on longer timelines toward goals that existed only in our heads and to work in larger groups beyond our touch and kin. Such a capacity is precious—and we take it for granted at our own risk.

So what does it mean to read and write fiction in the midst of a global crisis? What can literature and criticism possibly do to bring us closer to a more sustainable and just postcarbon economy? Or to put it differently: how might the humanities *underwrite*, as it were, a postcoal, postoil, and postmethane subjectivity?

The answer is not singular. It will not come from one of us. But it requires individual exertions, things to throw into the ring. A cooler future will need its literature and its literary critics.

An Energy Canon versus an Energy Unconscious

One task at hand is to build an energy canon.

That project, as critic Graeme MacDonald suggests, has neither proceeded very far nor resolved basic questions fundamental to its formation. Noting Amitav Ghosh's puzzlement over the absence of oil in American literature, MacDonald asks us to pause and consider "what constitutes *oil literature* in the first place?" What characteristics invite a text into the oil canon? "Must a work," he writes, "explicitly concern itself with features immediate to the oil industry?"[2] Thus far, most critics have assumed that an energy canon should have that quality and subsequently concluded that we don't have much literature to choose from. For instance, we hear in an oft-repeated paraphrasing of Ghosh that we lack oil classics simply because our oil culture resists being narrativized—that it is too quotidian, too saturate, too omnipresent, or perhaps too shameful and embarrassing to induce the artist to write.[3] The same might be said of our energy infrastructure beyond oil.

Nonetheless, a new generation of climate critics has tried to fill out a shelf of energy classics. *Cities of Salt*, a quintet written by Abdul Rahman Munif, appears, for instance, on just about everyone's list—a first step in identifying a global oil canon in the deserts of Saudi Arabia. Similarly, science fiction texts classified as cli-fi, most notably those of Kim Stanley Robinson, appear in this emergent oil canon because they ask pointed questions about technology, dystopia, and energy limits. Interestingly enough, even a canonical work like Herman Melville's *Moby-Dick* has been reclassified as part of an oil canon, as a text concerned with resource extraction that portends the demented history of our later petroleum culture. This book-by-book construction of an oil canon reflects a conscientious effort to make syllabuses speak to our most pressing political issues, but it also comes off as a little desperate when we scramble to resuscitate texts like Upton Sinclair's awful novel *Oil!* simply because it fits the bill.

The fact is we probably don't need an oil canon, or for that matter, an energy canon. What we really need are new methods for unpacking the energy unconscious that already sits behind the texts we read. MacDonald arrives at this conclusion toward the end of his manifesto. After he nominates a handful of oil texts to consider for inclusion in such a canon, he concludes that the real task is to learn how to see energy's presence in our literature even when it is "not in the foreground"—to recognize that everything from *The Great Gatsby* to *The Connections* is already in a sense an oil text, wherein energy sits there, just under the surface, "untapped, bubbling, . . . ready to be extracted."[4] This notion of an energy unconscious, initially articulated by critic Patricia Yaeger, has attracted a growing chorus of writers queued to that task, and it finds institutional expression in the collective statement made by the 101 authors of the volume *Fueling Culture: 101 Words for Energy and Environment*. To quote the volume's coeditor Jennifer Wenzel, together they have sought out new "protocols of reading and modes of inquiry that can perceive the pressure that energy exerts on culture, even and especially when energy is not-said: invisible, erased, elided."[5]

What drives this interest in the energy unconscious is the growing recognition by literary critics concerned about climate change that the problem we face today traces back to deep (and unrecognized) habits of combustion that structure our work and play, our thought and sleep, our rituals, artifacts, moods, surroundings, and ideas—habits that are now pointing us down a perilous path. Thus, to read literature in light of climate change means answering some difficult questions about where energy sits in our texts and our lives, such as what the hidden, if concrete, modalities are by which fossil fuels structure the potential for life in our literature and culture, what the coordinates (to steal a term from Yaeger) are through which coal, oil, and natural gas enable and constrain modern subjectivities in our texts, and, to return to where we started, how fossil fuels underwrite the constitution of a modern soul by fueling its well-being, its security, its fears, and its articulations.

We have bits and pieces coming together on these questions.[6] But one small confident step is simply to identify—a bit more systematically—where energy is and what it does in our literature.

Proposition 1: Ambient Energy, or the Presence (and Absence) of Fuel and Fire

The easiest place to start is with the presence of fuel and fire, and, more generally, with the ambient energy that surrounds us.

It is difficult to live without carbon. The cave without the fire is an uncomfortable place. Today, networked to the grid, many of us can take fuel for granted, but if calories are not being burned nearby, if fuel is not being gathered and combusted, it can get cold, dark, and uncomfortable very quickly. To understand that, all you have to do is turn off the furnace, the water heater, and the lights in the midst of a cold Michigan winter. The pipes crack, the air becomes frigid, the warm quilt turns as cold as the air surrounding it, and the nights sit wickedly black. The polar vortex is, in other words, a threat that we must plan for. The converse is also true. If the sun's fusion can't be shielded against in the heat of the day, if fans and air conditioners give out, if cool water can't be pumped in from afar to quench overheated bodies, then life can be a very hot, sun-drenched, and uncomfortable slog. India's heat waves in recent years, one of which saw temperatures climb to 125 degrees, and California's extended droughts are also the new disquieting norm.

The manipulated environment is, in other words, the pretext to life. In the North that life was built on the furnace, in the South on climate control. But both depend on a reliable infrastructure of energy flows, even while the fuel burns unseen at some remote power plant miles down the road.

To read literature in light of climate change means, first and foremost, heeling more closely to fuel—to the presence, and absence, of carbon's blaze in the ambient environment. It means seeing the acts of combustion that help to keep the world's volatility at bay and that mold the natural world to suit our bodies and temperaments.

1835. The White Mountains. A Cabin.

It was not long ago that the United States turned to coal.

Sometime before that, Nathaniel Hawthorne imagined a New England family sitting around a fire in the midst of a bone-chilling storm. They sit high in the White Mountains more exposed to the elements. The fire roars a broad blaze that lights up the room, its energy taken from the driftwood of mountain streams, dry cones of pine, and the great splintered ruins of trees. These gathered fuels, Hawthorne tells us, crackled and burned, radiating their heat and sending small emissions skyward in what was a universal ritual summoning to itself the forest and field and the sociality of families and travelers.[7]

Hawthorne sets this scene in a notch in the White Mountains at a time when the hearth—that focus of heat and camaraderie—still stood at the center of social life. He describes the wind outside howling and throwing its force against teamsters working to carry their heavy burdens through a mountain pass, men walking alone against a driving wind in search of themselves, and families holed up against the untamed energy of lightning and gales that send mountain rocks hurling. Nature's undomesticated energy— the sun sopping up ocean water, clouds churning, thunder and lightning, churlish winds—is the background to this family and this traveler huddling around the hearth.

"Ah, this fire is the right thing!" we hear, as a father with his "frame of strength" (and a mother of "subdued and careful mien") hurls pine branches onto it until the "dry leaves crackle," until the body is warmed, and the light "hovers" and "caresses" them. This is a "cheerful fire," a bit of civilization to stave off the gloom and all of those natural elements that do not fall under human control.[8]

Weather is the foreground here, climate the backdrop, and human agency is signified by people encircled around a small bit of fuel that signifies the human administration of culture.

1906. Chicago. Back of the Yards.

Even after the West's switch to coal, fuel could be scarce, and its scarcity lent a chill to the background of working-class life.

Upton Sinclair trains his eye on a Lithuanian immigrant staggering home from work through snowdrifts with "a bag of coal on his shoulders"—an image evoking the direct correlation between fuel and survival. The man opens the door to a thin house into which Chicago's famously cold winds blast. In a room, his wife and children sit "huddled" around a stove too puny to warm their bodies. The man lights up a pipe to generate the extra impression of warmth. But as night comes, because coal is expensive, the stove is put out and his children crowd into a single bed beside leaky weather boards, clothes and overcoats on, jostling each other to find enough body heat to get through the night.[9]

The elements are always raw, even when we don't feel them. But for working families in places like Chicago's Back of the Yards, the elements always had an extra menace to them. The world's energy infrastructure did not extend to the homes of the working poor when Sinclair was writing; life was tied more nakedly to the coal cart and to the pennies of wages needed to purchase that fuel. The privileges of combustion have always been uneven, and in Sinclair's text we see that inequality played out in countless little ways: in a small child struggling to warm his "naked little fingers in the unheated cellar," in the conversation of families worrying that they might "freeze to death," and in the "chill winds" and "shortened days" that bring risk to laboring bodies outdoors.[10] Industrialization might have required over 460 million tons of coal a year in the United States when Sinclair was writing, but coal for the industrial poor was a different matter.[11] It was not, and still is not, distributed equally. To them, coal was precious, and illness, draft, and damp the background to life.

The polar chills of climate change carry this social inequality forward. Keeping that volatility at bay will be more difficult for the world's poor.

1952. New York City. Below the Streets.

By midcentury, an infrastructure of electricity and natural gas networked the West's cities, suburbs, and much of the countryside, providing heat and lighting, but there were, as there always are, dark spots.

Ralph Ellison carries us to a basement in the most electrified city in the world, New York City, somewhere not far from Broadway, below the city lights where people still live in the shadows. There he imagines a man, an African American man descending beneath the streets to an underground chamber of electric lights to a home he has made for himself, a "warm hole . . . full of light." A squatter in the basement of a segregated apartment building, he has found a way to network himself to the grid, to pilfer a little energy from the Monopolated Light and Power Company so as to heat and brighten his room below the streets. He's constructed for himself a little warmth; he's wired the ceiling, "every inch of it," with 1,369 incandescent lights and has plans to wire the walls and, possibly, someday even the floor. The authorities know that "current is disappearing" into the jungle of Harlem, but they can't track the seepage. And so he lives on underneath modernity, warm—with bright lights blazing.[12]

The ambient energy of modernity pulses through our texts as heat, light, ventilation, refrigeration, and even music. But Ellison takes us to where this energy infrastructure, this otherwise immanence, can't be assumed—to the lives of America's racial minorities for whom that infrastructure came late and, in a few cases, not at all, and for whom the bills to support it could be prohibitively expensive. Ellison's electric lights, this metaphor for the thing that "confirms . . . reality, gives birth to . . . form," sits dialectically opposite the blackout, evoking the limits to modernity's imaginary, turning us to the dark energy that we don't see, this policing of modernity's comforts, its heat and light, along differential axes of power.[13]

Climate change returns us to this racialized energy that distributes carbon's privileges unevenly.

1996. Phoenix. The Postwar Suburbs.

Not so long ago HVAC systems reconditioned life near the world's hot zones, reinventing in the United States what came to be called the Sunbelt through a migration of refrigerated bodies.

The late contemporary writer David Foster Wallace imagined sitting indoors in Phoenix, a safe short distance from the uncontrolled heat that defines life in and near the world's torrid zones. Outside, the sun, he wrote, is "a hammer" producing a "fat . . . heat". But inside, the body is safely networked to an infrastructure of carbon that pushes that unmastered and unwanted heat to the background. The "spidered light of an Arizona noon" comes piercing through the glass. But characters don't feel it: they move from office to hospital to apartment in clearly laid "vectors from air conditioning to air conditioning," through interiorities "double-windowed" against that fat heat.[14]

Climate can feel residual here, something controlled and controllable—a ray of light that gleans through the window—in contemporary life. It can appear to be confined to the outside of the picture. But climate remains an issue, even if only implied. The movement from air-conditioning to air-conditioning is predicated on this high-energy structure of interiority that delimits the body's experience of the world. Climate has, in short, been resigned to, not so much controlled. For the time, we have cordoned off the sun. But in the Sunbelt, we live amid this sort of parallel huddling against the storm, with our cooling fires heating up the planet a little more each day.

Refrigeration is the irony of global warming: combustion everywhere, but no longer visible, confined to the cool ventilation of the modern interior.

Proposition 2: Propulsive Energy, or the Presence (and Absence) of Disembodied Labor

A second place to look for the energy unconscious is in the propulsive energy that appears in our texts; that is, in the kinetic energies of the machine which are the wages of combustion.

Life requires work, mobility, labor (whatever you want to call it), and energy is its presupposition. To get anything done—even an act of leisure—requires fuel and exertion, calories and labor, whether the subtle issuance of breath from a pair of whispering lips or the heft of a machine press stamping a piece of sheet metal.

Modernity proposes that it is an automated thing, sweatless, as if real bodies, manual laboring, had been removed from the equation, pushed to the fringes, and replaced by robots, tractors, cranes, and railways. The images are familiar: the backhoe digging a trench through California's hard clay without dropping sweat, the boom crane hovering over the city commandeered by a single operator, the container ship coming to port loaded down from across the Pacific without galley slaves, and the 250-horsepower engine hurling bodies uphill, downhill, and over mountains without burning edible calories. These propulsive energies can be subtle too: heard in the hiss of water pressure from a shower head or the whirring of a Whirlpool washer.

To read literature in light of climate change means feeling the propulsive energies of modernity again, tuning into its great labor subsidy that multiplies and accelerates the body.

1855 New England. A Paper Mill.

In its first act, propulsive energy asserted itself in the textile mills where waterwheels—drawing on the gravitational energy from mountain streams—allowed the West an early trial run at the factory system.

"Come first and see the water wheel," a young boy says to a client just up from the city. The boy works for a paper factory high up in the bleak hills of New England. The visitor, a character from one of Herman Melville's novellas, has traveled through January's bluster to inspect this modernized mill from which he plans to purchase the paper for his business in town. Entering the building the traveler is less impressed by the factory's output than unnerved by the mechanized processes that he sees. Life here seems to center on what he calls a "dark, colossal water-wheel" moving to a "grim" and "immutable purpose." It has called to itself rows of young girls, all laboring to the rhythm of its automated press—driven by disembodied energies that have no regard for the body's limits or the mind's tolerance for repetition. This first glimpse of the robot, of mechanized labor is forbidding. Melville imagines that it asserts itself with a "metallic necessity," that it has to it a certain "unbudging fatality" and "monotony" that takes the casual observer by surprise.[15] These energies, commanded by capital, are new, incipient, localized. Without coal's expansiveness, they have not yet become jejune.

Melville's nightmare, this "Tartarus of Maids," speaks both to the preposterous output that came with automation—these "piles of moist, warm sheets," dropping, dropping, dropping, from the press without end—and to the displacements of human energy that new work processes required. The observer here is left stunned, as well as melancholic, over the costs of recalibrating the body to the tempo of nature's energies, to the sight of these young girls losing their form, their volition, to repetitive folding. We are left with the haunting image of these "rows of blank-looking girls, with blank, white folders in their blank hands, all blankly folding blank paper"—these "cogs in the machine" that move "mutely and cringingly" to an inanimate force greater than themselves.[16]

Climate is ancillary to this story—a background to the energy shift to come—but we see in these churning mills a nearly equal opportunity for exploiting bodies and dementing nature's balances under the power of renewables. We see an early version of ourselves.

1909. The Industrial City. A Tramcar.

Propulsive energies appear first in industrial mills driven by waterpower then by a complex of coal and steam. But in their second act those energies show up on the road as the great circulator of goods, bodies, and resources.

D. H. Lawrence takes us to a moment when this new circulation was picking up momentum. He imagines the body being hurled through time and space on the most "dangerous tram-car" in England powered by compressed steam, boiling water, fired carbon—those kinetic energies that drive us forward. Lawrence settles in on the image of a passenger leaving the country, sitting in a calm "green and creamy-colored" tramcar, relaxing at the station to the train's subtle movements, pausing and purring to a rhythm that feels in keeping with the fresh air and quiet of rural life. But that lulling impression turns out to be premature. Soon the car is lurching and jerking through space—taking "reckless swoops," "bouncing" through loops, and "breathlessly slithering" toward an ugly and dreary city that looms in the distance. The train mutates into the reckless propulsion that is driving us into the gloom of industrialization, into this life of "fat gas works," "narrow factories," and "sordid streets."[17]

Lawrence saw in the radicalized horsepower of the railway, in its excess, its supplementarity, a certain derangement. This jerky little tramcar becomes the manifestation of a great disruption deep in the registers of modernization—a frenetic circulation without a center. Darting from town to town at a speed beyond horse or mule, we enter a topsy-turvy world, spinning in crisis, outside of its normative bounds—a world crammed with (what Lawrence portrays with characteristic discrimination and sexism) "cripples and hunchbacks," "howling colliers," and "fearless young hussies," who represent a grotesque distortion of humanity[18] This uncanny excess, this amassing of humans, this perpetual motion, is the catalyst for a set of corresponding social disruptions. Coal fires us here into the uncontrolled admixture and cauldron of excitation that blurs the old relations between city and country, time and space, class and gender, as brazen young women break gender conventions, unloosing their furies on men, blackened coal miners howl obscenities at the world, and the world whirls outside of its normative form.

Climate change points backward to this creative destruction, to the release of pent-up libidinal energies and blocked desires in the traditional world, but it also propels us forward into the wreckage piling up not far out of sight.

1940. Oklahoma. A Farm.

This excess of kinetic energy, administered by culture, came to saturate, even penetrate, life, bit by bit; but nowhere did its signature appear so clearly as it did on the farm.

John Steinbeck returns us to the dust bowl, to a time when the traction engine began to multiply and displace human energies on the farm. Settlers, tenant farmers, and squatters still struggle here, on the southern Plains, to cut life out of the land by hand. But the pressures are piling up. The flux of new technologies, and their bare living in the midst of a changing climate, keep that world on the edge. The tractor becomes the symbol of this disruption. Steinbeck writes that the tractors came "over the roads and into the fields, great crawlers moving like insects," "snub-nosed monsters," obeying a different logic than did the farmer or the mule.[19] Possessing energies that exceeded those of horse and mule they ran over hill and valley, gulch, and stone, through fences and courtyards, tearing up the sturdy mat of grass and flattening the West's soil as if by implacable force.

This propulsive work subsidy altered the fabric of things on the farm. Not only was the output far beyond the capacity of horses and men—elevators full of grain, shiploads of cotton—but so too were the dislocations beyond expectation, radical changes that registered in the sinews of rural people's consciousness. "One man on a tractor," Steinbeck writes, could overnight take "the place of twelve or fourteen families." He imagined that a great "iron seat," sitting high above the ground, had come between us and the soil. The new hired hand—gloved, goggled, and donned with rubber dust mask, straddling his tank of oil—was simply a "part of the monster, a robot in the seat," "muzzled" and "goggled" against the world, a laborer who no longer knew "the land as it smelled."[20]

The coming of dust clouds—climate's retort—was the cost of this great forgetting. It was an early warning sign of a combustion that ignores the fragility of life and co-opts the purpose of people.

Post-WWII. The Laguna Reservation. On the Road.

The world's propulsive energy never fully colonized American lives; it arrived in different sizes and distributions and was always parceled out, even rationed out, to different peoples for different purposes.

On the Laguna Pueblo, after the war, life registered the disembodied energies of modernity, namely, the pickup truck and the tractor, but as a lower-energy society, it persisted with the old embodied ways of moving and breathing. Leslie Marmon Silko gives us the image of a veteran returning from WWII, traumatized by its violence, explosiveness, and distortions. She peers in on him lying in a Los Angeles hospital on an "old iron bed," staring at the walls, unable to discern the past from the present, his memories linked like a train of colts, "one colt tied to the tail of a colt in front of it." Against the "posters of tanks and marching soldiers" that recall war's trauma are the quieter memories of home, of living bodies and living labor on the reservation—of "the creamy sorrel, the bright red bay, and the gray roan—their slick summer coats reflecting the sunlight as it came up from behind the yellow mesas."[21] Jumbled, her protagonist stumbles out of the hospital and onto city streets, falls flat to the ground, pressed to cold concrete, before being shuttled by train back to the reservation where there is some hope.

Silko addresses the melancholy and injury generated by modernity's propulsiveness. We see in her writing this internalized hostility—symbolized by uranium mines, atomic bombs, and images of war. In the context of imperial violence, of this great crippling, she asks, what can an Indian ceremony do? What can one person do to deprogram bodies keyed to the militarization of modernity's energies? In Silko's novel, modernity's propulsion appears as destructive or as having receded into the background. Reduced to bit parts. Uranium mines. A tractor stolen by a young man and ridden into town for alcohol. A GMC pickup truck crushed around its passengers like a "shiny metal coffin." Army jeeps. Rubbing against the grain of this white modernity, she calls us to the reservation's yellow mesas and to its living bodies eking out life amid a hostile climate—something meager, thin, surviving, but nonetheless something. Life persists here not so much in bright lights or imperial bombast but in the stability of place and ceremony—in the small acts of "poking kindling into the potbellied stove," in sharing "pieces of fried bread," and in the "broken shadows of tamaric and river willow."[22] To return ourselves to fullness will be a long road, she tells us, one that will require an engagement with the body, sweat, fatigue, and patience before we are able to move again deliberately on our "own two feet"—to return to the ears, the eyes, the body, and the soul.

Climate change means searching through the destruction, this imperial violence, for what is still there to recover.

Proposition 3: Congealed Energy, or the Deep Energy of the Exosomatic Environment

The energy unconscious is also fed by a third channel—by the congealed energy that manifests as steel, cement, glass, and aluminum arising around us and encircling our lives.

A sublimated form of fossil fuels, this energy is harder to see at first glance. If, on the one hand, we feel carbon's oomph in the stick shift of an automobile and sense its presence in the ambient energy pushed out of our HVAC systems, it takes some small extrapolation of the intellect to also see it manifest in our surrounding material culture—all of this deep heat, dark energy, congealed matter, manufactured in the furnaces of the world's steel mills, cement factories, and glassworks.

Congealed energy defines the literal infrastructure of modern life—its hardened surfaces, its walls, parking lots, and transparency. The world's furnaces have burned longer and at a higher pitch since we swapped out limited charcoal for seemingly limitless coal. Breweries, glassblowers, brickmakers, and ironworks scaled up production after the chokehold on fuel was taken off. Although it is hard to know what percentage of fossil combustion goes into the world's furnaces (rather than, say, its engines), the number is impressive. It likely hovers around 21% in the postindustrial United States and perhaps as high as 50% in industrial China where much of the world's hot and dirty furnaces are sited.[23] That heat allows for the verticality, the hardness, and the uniformity of modernity's material surroundings.

To read literature in light of climate change means tuning into this deep matter of the exosomatic environment that rises like a skyscraper and spreads out like a parking lot around us.

1860. London. A Blacksmith's Shop.

Fuel was once scarce and metal precious. We did not always live surrounded by steel, glass, and concrete.

Charles Dickens takes us to a time before the energy deepening, when iron was handcrafted and doled out in smaller rations, when the city had yet to rise up on steel girders. He brings us into the home of a blacksmith, where a squadron of London's soldiers have just arrived in chase of a felon. They require fire, a white-hot heat, from the blacksmith's forge to remake broken iron manacles. He takes us back into the blacksmith's workshop, the little world of the artisan that has an intimate texture to it compared to the future steel mill, evoked by roaring bellows, red sparks flying, hammers and anvils clinking, and the proximity of the forge to the kitchen.[24] We have yet to come into the ubiquity of steel, glass, and mirror: metal is still a craft, and there is no steel wall, no wall of concrete, no sheet glass, rising between us and the world.

Dickens wrote at a time when coal was delivered by cart, and when the world's furnaces were small and localized, fuel burned on site. To be sure, coal was everywhere in Dickens's world, because Britain's shortage of wood, its island status, demanded an early conversion to coal. But the industrial furnace had not yet delivered this dark matter in staggering mass. We might think of London's Crystal Palace, this national symbol of a world constructed out of sheet glass and cast iron, as the future to come—as something not yet arrived. For the time being, the material culture of the living poor was more modest in terms of congealed energy: glassware and ceramics remained rare purchases, bricks costly, and metals reserved for utilitarian functions like stoves, clasps, pots, and nails. In Dickens's writing, iron's presence shows up only sparingly, and it has a class texture to it: "light iron" rails on the staircases of the wealthy, "iron bars" and "iron gates" on the windows and grounds of a rich woman's house, "iron safes" to hold the savings of those people fortunate enough to have them, and symbolic "iron legs" put on the less fortunate, these fetters of constraint. The sound of metal is only pleasant here when we see the artisan at work and when we hear a mare passing in her iron shoes. More representative is the recurring image of a convict in chains "filing at his iron like a madman."[25]

Every blast of the forge heats up the planet, whether small or large, and so it is our job to know what the social purpose of that heat is.

1900. Chicago. A First Look.

Sometime in the nineteenth century, a windfall of cheap carbon produced a modernized space of verticality, hardness, opacity, glimmer, and transparency.

It is easy to miss the novelty of the urban environment today, at a time when more than half of the world lives in cities. But the stunning and overwhelming impression upon first seeing the modern city, this rising exosomatic environment built on congealed energies, can be felt in countless vignettes in our literature. Theodore Dreiser provides us with one such glimpse in his portrait of a young woman arriving in Chicago, this city of over half of a million people, after a train trip from the country and its rolling prairies. She looks up to see the city rising skyward on brick, stone, and glass and spreading out endlessly on pavement. "These vast buildings," she thinks to herself, somewhat dazzled by the enormity of it all, "what were they?" What was the meaning in these "large plates of window glass," "frosted glass," "miles and miles of streets and sewers," and "immense trundling cranes of wood and steel"? She senses that they signify "strange energies and huge interests"—a "power and force she d[oes] not understand." The outsized proportions of this world—the sight of all of this asphalt pavement, plate glass, baked bricks, and steel—is hard to get one's mind around, and it feels to her less graspable than the "meaning of a little stone-cutter's yard . . . carving little pieces of marble for individual use."[26] Something exciting and terrifying has happened.

Dreiser's city evokes in its images the deep heat of modernity, the hidden work done by blast furnaces and coal yards to rebuild the world around us. What that world means points to at least two different messages in his text. First, its rising buildings and frosted glass are a metaphor for the perception of progress, for something "wonderful," of "importance," "prosperous," something that speaks for "success." That is, this building-ness represents the idea of an actualized movement in time and space toward something better. But as importantly, this modernized space is also one of exerted power and its corollary, helplessness. To this young working woman, it seems to have a personality, a "high and mighty air," that gives her the feeling that it is "calculated to overawe and abash the common applicant." In other words, the city's built environment emphasizes not simply progress and accomplishment but the insignificance of some people, those who are not dressed in "fine clothes" and not invited through the door. That exclusiveness is spatial and classed—as the protagonist leaves behind these imposing edifices and passes into "a region of lessening importance . . . until [the city] deteriorated into a mass of shanties and coal-yards."[27] These congealed energies are modernity's literal stuff of power and exclusion.

Today's climate calls us back to a public and privatized material environment that heats up the planet while opening itself to some of us and closing itself off to others.

1939. New York City. A Construction Site.

Modernity's exosomatic environment arises out of deep energies that are not always obvious. Mortar, brick, and concrete—these are the quotidian materials of modern life, the congealed energies, the stuff of the present.

The immigrant writer Pietro di Donato takes us to the scene of an Italian bricklayer climbing up the scaffolding of a new building in New York City, hovering on a rickety floor somewhere above the city streets. This is the exosomatic environment of modernity composed of energy-dense building blocks—brick, made of country clay, and shale fiercely heated in furnaces at two thousand degrees, compressed, vitrified, and trucked into the city. We call it the age of steel, the age of coal, but it might as well be the age of cement, brick, glass, the deep energy of modernity. Di Donato's novel brings us close to the hard rock and pooled cement that beats against flesh in the city. Here the world feels different: the touch of "stoney cement pudding," of "warm plushy soon-to-be stone."[28]

Congealed energy asserts itself in the modern environment in the texture of living, the shrill verticality of the city, the hardness of labor. Di Donato imagines that these buildings want "to tell [us] something"—that they speak and breathe with modernity's heft, infused with a bit of living flesh.[29] His story focuses on the lives of struggling construction workers, sweating above the city, laying out their lives brick by brick literally, in the heat and chill of the day. These are men who scrape against the city environment and who leave a little of themselves in each slap of mortar, each clump of brick. He calls them men "born in a mortar tub," "bodily machines" whose "human water commingles with lime-mortar and brick."[30] To them the city is not just a landscape—seen at a distance, a skyscraper against the clouds—it is something lived, embodied, felt. Every urbanite knows this city, but these construction workers symbolize a hardened world that absorbs some of us in its making. Di Donato offers the chilling metaphor of a "Christ in concrete," a bricklayer crushed by a fallen building—crucified on a steel beam, buried in cement, his flesh embedded deep in modernity.

Climate today heats up the hardened city, it wears down bodies, and this congealed energy, this dense matter of modernity, demands more fuel as it absorbs us.

1985. Somewhere in Texas. On Horseback.

As this energy deepening becomes increasingly blasé, the energy unconscious convolutes on itself, manifesting as nostalgia, melancholy, fear, and fantasies of stripped-down environments, of a naked, unprocessed world.

We might also return to, in other words, where coal is not, to where oil is gone—to where congealed energies are absent. Cormac McCarthy takes us back into a bare, almost naked, sunburned landscape, an imagined nineteenth-century frontier world with only some desert scrub, a few ocotillo, and not much else. Fuel is scarce here, bodies the main prime movers, and the natural environment austere. The cowboy, the filibusterer, the hunter, and these wide, empty plains make their reappearance, that is, deep in a century of steel, glass, and concrete. McCarthy's characters are a dialectical product with an antimodern profile. They are the descendants of "hewers of wood and drawers of water," they have worked in "sawmills," they are "harness makers," and they float in and out of earthen villages, wooden shacks, mud-brick churches, stockyards made of wood. Their horizon, unlike that of their city counterparts, is unobstructed by buildings, always visible, and the ground made of dirt, mud, water, rather than pavement, terra firma felt through leather shoes, boots, and bare feet. We have left behind steel, glass, and cement for a bare landscape defined by "dark little archipelagos of cloud and [a] vast world of sand and scrub shearing upward into the shoreless void."[31]

For McCarthy, this stripped down world of scalp hunters, unnamed boys, and bony horses becomes the setting for an apocalypse of decay, a reissuing of the past to signify a future to which we are pointed. The congealed energies of the furnace are distant in this text, only showing up in fragments here and there, or else they have disappeared altogether. A Toledo sword, handcrafted, hammered steel. Pistol balls. Metal knives. Bottles of wine. An iron gate. An occasional mirror over the counter in a tavern. The materiality of McCarthy's world is that of the low-energy landscape where fuel is hard to find and where the built environment manifests in impoverished forms: as a "crude hut" of mud and reeds, a "dried cowhide door," "shoeless mules," and surroundings of an earthen form of baked clay, hewn lumber, and a few nails. This is the future as apocalypse—barely built, barely constructed, without the furnace, without fuel.[32]

Climate change requires digging into what manifests as fascination and nostalgia for this more naked and unprocessed natural world.

Proposition 4: Polymerized Energy, or the Textures of Modern Life

Fossil fuels enter into the energy unconscious of our lives through a fourth channel—in the polymerized energy derived from petroleum and gas feedstock that produces a global consumer culture of synthetic fibers and other carbon derivatives. This is not the deep heat of modernity but its fibrous and oily textures.

Nylon stockings, PVC pipes, asphalt roads, Patagonia rain jackets, Prana yoga pants, Montgomery Ward's polyblend nonwrinkle shirts, hairspray, automobile interiors, plastic chairs, laminates, synthetic oils, plywood resins, epoxies, fiberglass. The modern world is not just a heated thing of deep energy, of congealed matter; it is also a polymerized world, compounded and synthesized from hydrocarbons and mixed with nature's fibers, resins, and other organic matter. Petroleum and natural gas became in the twentieth century the feedstock for remolding the world, giving it a different quality, texture, and smell and a different backstory. We came to live "fractionated lives," Matthew Huber says, as we came into new selves distilled in the refinery and compounded in the chemical plant.[33]

This is not unfamiliar. We know that polymerized world by touch, its metonym being the plastic bag tugged from the produce aisle in a grocery store or found floating in the ocean, a commodity that stands in for a long chain of synthetic commodities and an equally long referent chain that sees in those synthetics the artifice, disposability, and inauthenticity of the modern world. But there is a flipside to that story, as Roland Barthes observed awhile back. Plastics and other fractionalized products from hydrocarbons also give to modern life its emancipatory quality—what Barthes called the "very idea of infinite transformation," or, in one of his more poignant phrases, "the euphoria of prestigious free-wheeling through Nature."[34] Polymerized energies, he reminds us, revolutionized the world's mold—tying life to the oil well, the bitumen pit, the fracking well, and the chemical plant.

To read literature in light of climate change means grappling with how this polymerized energy recast our world, how it gave to it a different elasticity by remolding the body and consciousness around a flexibility, a roundness, a cheerfulness that traces back to the harsh extraction practices and cordoned-off refineries of the world.

1852. The Cotton South. A Plantation.

Before hydrocarbon cracking retextured life, cotton was king and nature's fibers more assertive—leather, thatch, wood, silk, each the stuff of living.

Harriet Beecher Stowe takes us back into the cotton South, to a Louisiana plantation, where cotton, wool, and silk define the intimate (and violent) textures of life. She drops us in on a gang of slaves, embodied labor, racialized and sweating, working to clothe the West, to cull these pliant fibers from the soil. We watch an overseer, high on his horse, hovering over several slaves who are picking cotton to meet their quota. A northern slave, punished and unrepentant, arrives in the fields, new to the picture. He has been sent south for his transgressions to sow and harvest this fabric of life. To him, this little corner of the United States, which seems to be charged with growing the entire world's fibers, is an uncanny space, raw, vicious and disquieting, comprised of "wearied" bodies laboring in the "heat and hurry" of the season, families living spartanly in little "shanties," and "hoarse, guttural voices" reflecting a manualized world of grinding corn, picking cotton, sweating, repetition, and ache. For now, before he breaks free, he tramps day by day with everyone else as part of a "weary train," dragging bags and carrying baskets filled with raw cotton bolls that will soon clothe more privileged bodies, drape from their four-post beds, and decorate their dining rooms.[35]

Stowe's world has no plastics, none of this polymerized energy; it also has little iron. It is a type of world that we tend to call, for lack of a better word, natural. From the coast of Georgia through eastern Texas, that world was for over a century "piled with cotton-bales." Cotton could be seen sitting stacked at river docks awaiting transit, being dragged in canvas bags behind workers in the fields, bursting out of wicker baskets in storehouses, and waiting in the receiving rooms of textile factories across the Atlantic. Unprocessed cotton—this fibrous commodity—made up both the foreground and the background to life. In that world, we catch glimpses of slaves stealing away for a minute to find a nook among cotton bales, of visitors to the plantation resting against cotton bales, of owners and overseers laying down their drinks and whips on cotton bales. Cotton's eminence is so complete here that it even asserts itself on the muddled consciousness of a nation stuck in slavery in which even the good seem to have the mind of "a bale of cotton,—downy, soft, benevolently fuzzy and confused."[36]

Hydrocarbons have yet to enter the picture; polymers are for the future: fibers are still natural. But climate change requires theorizing how these hydrocarbons came to rewrite the arrangements we once had among soils, bodies, and commodities.

1874. Rural England. Out to Pasture.

A polymerized world came very late, well after modernization had produced in mass these congealed and propulsive energies seen in the Crystal Palace and the Corliss steam engine, but because the world's steam mills required raw unprocessed materials, wood and wool, modernity always had an organic quality to it that was never lost, even if it was easily forgotten.

Thomas Hardy returns us to Victorian England to a pastoral world that contrasts with that of modernity's bustling crowds, rigid factory system, and swollen furnaces of molten iron. Here is something different—the fleshy rather than mechanical world of shepherding, ranching, hunting, butchering, and shearing from which industry derived, among other things, its leather and mill wool. Hardy focuses in on a sturdy and reliable shepherd who spends his days in the fields tending to the world's sheep and to its supply of wool, far away, we are told, "from the madding crowd" and its industrial and mercantile bourgeoisie. No polymerized energies appear here, and machinery too is mostly absent. Instead, our protagonist lives surrounded by wool sheared from the flock and leather from cowhides. These more natural fibers texture his world, and they are the product of nurture and manual labor, of tending to the "new-born lamb," grinding shears for the annual fleecing, and walking among the animals with a "quiet energy" and a "steady swing" that speaks to this other textured life.[37]

For Hardy, the shepherd and his wooly business might be an allegory for a traditional Christian nurture, but modernity's textile mills—and commercialized wool—are nearby even if we never hear of them. The shearing we do see, this "soft cloud" of wool left behind at the end of the day, is really the beginning of a commodity chain that runs through markets and train tracks to the steam mills. Hardy permits us refuge in the fields, but these fibers are part of the operations of that world, even if there is an enduring innocence in the image of fleece falling from a sheep, these "three-and-a-half pounds of unadulterated warmth" that are meant for the "enjoyment of persons unknown and far away," people who, we are told, "will . . . never experience" this wool as it is, "new, pure," dropped from rams and ewes, "startled and shy." This facet of modernity's texture feels like it is tied up in "old habits and usages," but, to be sure, tending to these wooly beings was, and is, part of the warp and woof of modernity.[38]

Climate change means looking not just to the oil well and the polymer factory but also to the land where we derive our natural leathers and wools that even today drive the mills that texture modern life.

1966. Los Angeles. A Motel Room.

Polymerized energies from a feedstock of gas and petroleum entered into life, marking a radical break, in the middle of the last century when hydrocarbon cracking became commercially viable and chemical industries stepped in to stretch nature's malleability.

Thomas Pynchon takes us into this polymerized world of Tupperware parties, plastic TV cases, stretch slacks, vinyl records, plastic credit cards, blacktop parking lots, and "booze poured into cups of white, crushed, plastic foam." His world is, as he puts it, a "prefab" world of synthetic carpets, air fresheners, and air-conditioned motel rooms, where life is played out against synthetic textures. Pynchon portrays that world as a superficial one in which material goods and social relationships float, like loosely tied signifiers, across the page and plot and where life is symbolized by "the greenish dead eye of the TV tube."[39] Plastic is not the foreground in Pynchon's work nor is it the metaphor that holds the postindustrial and postmodern world together: it is rather that world's metonym, a recurring stage prop, that attests to its malleability, its ever-changing and slipping forms.

What we get in Pynchon's text is a world where the lack of fixity, its plasticity, its cracking open, its compounding, its atomizing, parallels not only the language of the text but the fractionalized products of the refinery, the hydrocarbon cracking of nature's once relatively stable forms. We see this great undoing in Pynchon's image of an aerosol can of hairspray, a little petroleum product, a polymerized substance made of volatile solvents and plasticizers, that a character bumps off a bathroom shelf in a cheap motel. She punctures it and releases its pent-up energy. For a moment, that unloosed can becomes the metaphor for a greater whirling and undoing of the world. "The can hit the floor," Pynchon writes, "something broke, and with a great outsurge of pressure the stuff commenced atomizing." This can of hairspray flying through the motel room, bouncing off walls and toilets, crashing into mirrors, and sending its synthetic "sticky miasma," its "fragrant lacquer," throughout it, works to signify these weirdly deep ways in which carbon products have retextured life and propelled us into something new. This great "woosh and buzzing" seems, at least to the novel's main character, to have "no end to it"—to go on without intent, until at last, depleted, it falls to the floor in a dizzying crescendo of exhausted energies, broken mirrors, and petroleum distillates.[40]

Climate change will require a better understanding of this polymerized world, compounded and molded around carbon's distillates, a plasticity that takes on a life of its own outside the laboratory.

1999. New York City. Inside a Waste Firm.

A polymerized world became nothing short of second nature by the end of the twentieth century—carbon's byproducts no less natural than the air and sunlight.

The reach of polymers spread quickly into life during the Cold War, when modernity's plastics and other petroleum products came to populate the foreground and the underground of people's lives. Don DeLillo takes us into the thick of an elastic and increasingly disposable society where our food appears as "plastic pouches" boiling on the stove, where street curbs lined with "identical black plastic bags" become part of the landscape, and where faces with puckered lips and painted eyelashes are "put together out of a thousand thermoplastic things." Plastic even comes to engulf our sexuality, manifesting in the feel of "plasticky" condoms, "polyurethane sheathes," worn by men in the throes of passion or by a boy who jerks off behind the "fiber-glass curtains" of a shed. What is left of the natural world is, moreover, fil-tered, as often as not, through plastic—through the tube of a TV screen or the "key stroke . . . mouse-click," the glow of a "plastic, silicone, and mylar" computer screen.[41] The texture of the world has changed: polymers have become normalized, a part of us. They have found their way into time and space, and into our bodies and rituals.

DeLillo sees in this polymerized artifice a metaphor for the emptiness of a petrol-fueled consumer culture, framed by fears of nuclear annihilation, that has lost its center, has become agitated and broken. Modernity's ener-gies are reduced in his text to forces that either cause injury or are trivial, and they are headed straight for the waste dump. Consumer goods, for in-stance, this *objet petit a* after which we chase, can be found "gleaming on store shelves" but they have lost their luster: they already appear to us, even before purchase, as garbage's potential, as waste waiting to happen. Moder-nity's representative becomes in his text a manager of a waste containment company, with his Tuesdays devoted to the recycling of "plastics, minus caps and lids," his life's work represented by conveyers belts of refuse, "four hun-dred tons a day," flying by with tin, paper, plastics, and Styrofoam, and his horizon a man-made landfill, shimmering with a "polyethylene skin, silvery blue," almost beautiful, like an "enormous gouged bowl, lined with artful plastic." In such a world, DeLillo wonders where things might end. Where is this polymerized world headed? Will it end in "some forgotten core of weary faithful huddled in the rain"?[42]

The structure of climate change has plasticity as one of its features; these polymerized energies stretch out the world's resources while at the same time accelerating us into an entropy that it can't quite contain, recycle, or deliver back into order.

Proposition 5: Embodied Energy, or Bioenergetics and the Population Spike

The fifth channel through which fossil fuels enter into our texts and lives traces back to the role they play in the bioenergetics of modern life, that is, in underwriting a global population boom.

To understand the energy unconscious is thus to come to grips with its deepest embodiment: not as coal dust on the miner's body, not as cement on the shoes of the bricklayer; not as emissions from a truck but as food, the marrow and milk, of living bodies—the chief input that makes it possible to grow modern populations, to gorge ourselves to obesity, and to see *Homo sapiens* sprout as a global dominant, a species so successful, so detached from earlier growth cycles, that it threatens its own survival. Today, as much as 35% to 40% of the earth's terrestrial plant growth is channeled to consumption by humans rather than other species, and we have become an ecological force capable of inducing a sixth extinction event.[43] Which is simply to say: fossil fuels also find themselves in our digestive tracts. The process by which that happens is, however, not straightforward. On the one hand, modern calorie production (which is the premise behind any population spike) is thoroughly new, rooted in tapping subterranean energy flows. The deep refrigeration, high-pressure heating, and methane needed to produce nitrogenous fertilizers are now the basis for reproducing life itself, for recycling depleted fields year in and year out to feed 40% of the world's population, which is substantially more than what nature's more humble nitrogen cycle can produce.[44] Similarly, petro-pesticides, herbicides, and fungicides are the key inputs for fragile hybrid crops that produce in spades but don't survive without a protective apparatus. And, of course, kinetic energies are needed to circulate grains and to pump water from underground aquifers. On the other hand, these technologies sit alongside the old way of doing things. Migrant men and women still work in rounds hand-picking America's crops, pouring sweat and ache equally deep into our food supply.

To read literature in light of climate change thus means attending to the presence of food and populations in our texts, to what we call demographics.

1870. North Dakota. A Settlement.

Carbon was slow to enter the bioenergetics of life. In its first stage, coal-fired railways integrated open soils and other isolate locales into a global food system by breaking free of earlier transportation limits.

The immigrant novelist Ole Rølvaag shifts our attention to Norwegian pioneers in the Dakotas, who, on the heels of the military conquest of Native lands, turned over its tough grasslands by hand, plow, and oxen and tied that soil piece by piece into the world's granaries. While we don't see or feel fossil fuels cropping up in the daily life of their subsistence settlements, we do catch glimpses of that energy appearing in the ligaments of market integration—in the train depot, in the market town, and in the lengthening of the railway track. Rølvaag captures carbon's creeping encroachment into our food systems by imagining the railway extending closer to the Dakota settlements. "One day," he writes, "a strange monster came writhing westward over the prairie . . . ; it was the greatest and the most memorable event that had yet happened in these parts." The Norwegians welcomed it, he writes, with "a joy that almost frightened them." It calls up for them fantasies of wealth and progress—of selling grain, of harvesting potatoes, and of thriving.[45] Rølvaag's train is the connective tissue for the world's modernized food supply.

What we see in this portrait of immigrant settlers is a combination of the new and the old. The railway is the symbol of the new, but the old is rooted in a fairly recognizable farming economy based on muscle power and the excitement of coming into "rich soil" new enough to produce a bumper crop for a few years. That story played out again and again across the world's frontiers in countless ways. But this rural economy was also an economy of precarity and limits, dependent on unpredictable rains rather than pumped irrigation and predicated on the accumulated nutrients of the Plains' rich loamy soil rather than petrochemical inputs. Climate's presence was fairly overt here, reflected in anxiety over drought, worry over too much rain, sleepless nights over untimely freezes. Rølvaag imagined that precarity in dying cattle and the limp body of a pioneer freezing in the snow.[46]

Climate is everything on the farm, and it begs us to learn to trace back through the connective tissues that produce the world's bumper crop.

1909. Chicago. The Board of Trade.

As railways integrated the hinterland and as steam ships integrated the waterways, oil and coal came to be mixed increasingly with human energy, producing food security and booming populations.

The pretext for this embodied energy were new and enormous economies of scale, massive mechanized grain elevators filled to the brim with wheat, driving populations skyward. Frank Norris drops us in on the Chicago Board of Trade, where trading futures in wheat, pork, and soy refers us back to the great hinterland of production where coal, nature, and capital produced this new bounty crop of calories. "Wheat, Nourisher of the Nations," he writes, "rolled gigantic and majestic in a vast flood from West to East . . . like a Niagara." Although we only see glimpses of carbon's presence here and there, it sits behind the scenes in these "ponderous freight cars," trains "reeking with fatigue," "great lakes steamers," "grain boats," "switches, semaphores," and "tower[ing] . . . hump-shouldered grain elevators." Coal, at the time Norris was writing, was the material basis for this "infinite, inexhaustible vitality" that made wheat into "a world-force," something "raging and wrathful," like an earthquake or a glacier. Carbon mixed with human energies fueled, that is, this incessant pounding in one's head: "Wheat—wheat—wheat, wheat—wheat—wheat."[47]

Norris's depiction of embodied energy, modernity's bumper crop, says little about the ecology behind it. Rather that energy is taken for granted; a background setting that has been absorbed by the market's pieces of paper and commodities, traded back and forth on the floor of Chicago's Pit. We see glimpses of actual wheat, these nourishing calories in the "scattered grain" left in the pits after the sale, in a few "paper bags, half full samples of grain," and in the "grain elevator" on the edge of the track, but wheat is transformed into a concept, a distant referent, a notice in the day's "crop report." It becomes the image of "waveless tides, . . . crushing down the price," or "eighty million bushels" being sent on their way to Europe. No longer do these sacks of wheat get carried by hand to the mill; no longer is the trade in embodied energy direct. Instead our food supply is reoriented around a fossil infrastructure that allows capital, miles away, to tug into "its undertow" the grain fields of the Northwest and to "ma[ke] itself felt" in the bellies of coal miners as far away as Prussia.[48]

The world's grain crop, this bulking up on food, is predicated on these carbon flows that have given to capital its enhanced ability to administer the farms of the world and to manage where climate change takes us.

1967. The Suburbs. At the Dinner Table.

A further energy deepening in our food supply came at midcentury with the spread of petrochemical inputs that produced the excess we know as the Green Revolution.

John Cheever takes us into the postwar American suburbs where food is fast and secure—and where the ornamental lawn has displaced the garden and the field. Food comes from the grocer by this point, and dinner parties, business dinners, and fund-raising dinners are central to the social fabric, central to performances of the self. Wafting from the suburban kitchen is the flavor of plenty: we smell a "leg of lamb," "frying bacon," "scrambled eggs and sausage." Petrochemicals are quiet, demure, here, showing up in cameos, in bit parts. But nonetheless they sit confidently behind the scenes. We hear them, for instance, in the background of the story's protagonist, who had worked for Monsanto and then spent another three years "analyzing chemical fertilizers" for the United Nation's Food and Agricultural Office, and we hear it again in a friend's lucrative investments in Merck, a leading chemical manufacturer of livestock insecticides, cattle vaccines, and an array of petrochemical products.[49] But otherwise they do not announce themselves or make themselves easily known.

The flipside of this food security—this normalization of a bounty of meat and fresh produce—is the excess that came with it in the suburbs. Cheever offers up a sympathetic caricature of a man who finds self-realization in food, an "absolute absorption."[50] He is an obese used-car dealer, weighing three hundred pounds, who likes "to buy food, cook food, eat food, and . . . drea[m] of joints of meat and buckets of shellfish." Consuming food takes center stage in this midcentury text, while its production disappears, shrunk down to the vague fear of "drought threaten[ing] the wheat crop," the sight of "wheat fields" out the window on a drive, and a nostalgia for the "smell of eucalyptus, maples, sweet grass, manure from a cow barn"—to a remembrance of rural things past, once "intimate, human and pleasant."[51]

Climate change requires seeing the origins of this bumper crop, its inequitable excess, and the oil that is being sowed into the lives we live.

The Future. Inner Horner. On the Edge of Security.

Oil, coal, and natural gas drove up the food supply, overriding nature's life cycles and taking us into a planetary carrying capacity bursting at the seams. George Saunders creates for us an allegory of this cramped world populated by a handful of brainy, math-calculating, alien beings called "Inner Hornerites" who can't physically fit within the confines of their own country, their territory so puny that only one of them can live in it at a time. An inventory of their natural resources includes one small apple tree, three cubic feet of cracked dirt, and a nearly dry stream. The country's excess population waits on a small plot of land granted to them by their bullying neighbors, a slice of land so thin that they stand body part to body part, sleeping, thinking, and waiting, "timidly," to get their chance to stand in their own country. Eking out this meager life, one Inner Hornerite floats the idea of a "hunger strike" to raise consciousness. "Excuse me, Gus," a friend says gently, "But aren't we sort of already on a hunger strike? Because no food? . . . I'm not sure how effective a hunger strike would be? I mean they might not even notice we're on one."[52] This trope of hunger and inequality reflects where some of us have already arrived today.

Fossil fuels, to be clear, are literally absent in Saunders's text. But they assert themselves in the contextual origins of the demographic crisis and the related crisis of humanitarian will faced by its characters. We see in Saunders's text an allegory of our planet's unequal distribution of resources in an environment in which population surges to preposterous proportions. The crisis might be looming, he seems to say, but those of us in the West continue to be expansive beings who eat heartily, think highly of ourselves, and move around a bit too freely. Saunders caricatures those of us living in the core as big coffee-drinking beings with "historically righteous bellies" who press hard on the cramped lives of the dispossessed. We can hear ourselves in the bombast of their emerging leader whose brain sits literally in the gutter, having fallen from his head: "I've been thinking about how God almighty gave us this beautiful sprawling land as a reward for how wonderful we are. We're big, we're energetic, we're generous, which is reflected in all of our myths."[53]

The West, this hungry and overfed empire, bears the moral brunt of the coming climate crisis, and it does us no favors to lack empathy for others' hunger and plight.

Proposition 6: Entropic Energy, or Waste and Dissipation

The sixth, and final, channel that feeds the energy unconscious in our texts derives from the entropic energy, the waste and dissipation, that follows by iron necessity from the act of combustion.

Waste, the squandering of energy, is the human condition. Entropy is a universal fact rather than some act of consent. Yet there is more to it than that. How we expend the energies surrounding us, whether dense finite ones like coal and oil or diffuse renewable ones like wind and water, falls within the realm of human agency. Life might produce waste; sustainability might only be an imperfect goal. But there is still plenty of room to act in between—endless opportunities for temperance, longevity, and benevolent outcomes.

The exhaustion of nature's energies is not an equal affair—not all acts are morally commensurate. For instance, on one end of the spectrum, we have the hostile exhaustion of energies experienced in the propulsive chain reaction that spent itself over Nagasaki and obliterated life for miles around: uranium atomized and expended, the dismemberment of life. In the middle, we have infinite personalized and impersonal acts of slow violence—the unintended disordering of life, symbolized by exhaust spewing from the world's one billion cars and trucks, emissions from a great hurrying that is undoing life. The hopeful end of the spectrum is represented by a tempered entropy, anchored in collective efforts to forestall, or to suspend, disorder for the time being—in little acts of resistance against things falling apart. This staying of disorder is seen in the seeding of a tree to capture and reorganize carbon, in the composting of food wastes to extend soil's life just that much longer, and in the collecting of solar radiation before it bounces back off into space and loses its use to us. Entropy might be the condition of life, but it is also life's business to try to stay it for awhile, to hold it off for the sake of ourselves and progeny.

To read literature in light of climate change thus means tending to this dissipation and loss in our texts, to the waste, emissions, and wreckage that go along with carbon's combustion.

1861. A Steel City. In the Iron Mills.

Entropy finds its most obvious signifier in the trail of smoke that follows industrialization wherever it appears. That trail, once visible from a smoke stack here and there, is now ubiquitous, and it increases in size by the day.

Rebecca Harding Davis wrote about this exhaustion at a time when the fossil economy was first scaling up. Smoke, the sign of coal's waste, emerged for her as the "idiosyncrasy" of industrial life, its chief trope, its metaphor, metonym, and synecdoche, the sign of a great disordering of things. The imagery she constructs has a hyperbolic aspect to it that is only a little exaggerated: smoke rolling in "slow folds" from great chimneys of iron works, grease coalescing in "black, slimy pools" on city streets, soot blackening "the house-front" and "the faces of the passers-by"—but everywhere smoke, "smoke everywhere." To Davis, this exhaustion saturates life in the iron mills, sinking into the grooves of the environment, its street corners, and into the pores of the working class, people who live surrounded by, in fact, almost absorbed by, ash heaps, foul smells, grease saturated air, and soot-covered ikons, plants, and birds, by this trace of combustion telling us that life is depleted, "almost worn out."[54]

Soot is the sign of literal waste in this novella. But it rallies to it a train of human wreckage and social waste, the loss of human energies, under the management of fossil capitalism. The images are explicit: coal-sooted faces "bent to the ground, sharpened . . .by pain," broken bodies "stooping all night over boiling caldrons," wheezing infants breathing "air saturated with fog and grease and soot." These are lives spent and stunted before they can grow. Davis offers up the tragedy of an artistic metal worker, who has carved out of korl, sculpted out of waste metal from the furnaces, figures "hideous, fantastic . . . but sometimes strangely beautiful" that communicate the passion and emotion of an artist whose life is incomplete, retarded, by this "vile, slimy" condition he inherits. We are left to contemplate this frightening sculpture made out of waste: a nude, muscular woman, grown coarse with labor, beautiful, starving, and with "arms flung out in some wild gesture of warning."[55]

The waste products of carbon—the depletion of fuel and of people—this eerie smoke and these beseeching bodies are written into the politics of climate change.

1920. Georgia. A Cane Field.

The waste of modernity's energy is both literal and metaphoric, with deep homologies between carbon's actual waste products and the human carnage fueled by the fossil economy's wasteful practices.

Jean Toomer takes us back to the American South in the 1920s, to Georgia's segregated world of sharecroppers and its sleepy fields that were out of step with modernity's bright lights. Liberating energies are glimpsed in the distance, in cities with "copper wires," "a powerhouse," "yellow globes gleaming on posts," and a bright "incandescence," but here is something darker and different: traces of pine smoke, a sawmill hugging the earth, mule-drawn buggies, and cane fields where "time and space have no meaning."[56] The expenditure of carbon's energy, signified by a train rumbling through the village six times a day and shaking the earth, does little to emancipate bodies here. Instead life centers on small forges hand packed with coal, cabin hearths of fire and ash, and black arms swaying with scythes. Fossil capital has not so much bypassed this place as it has left it in the periphery, rendering it a resource colony.

Wreckage is everywhere in Toomer's work. Smoke trailing from the shanty of a sunken woman living near the railway tracks. Burning flesh left by a mob and high-power search lights following the execution of a lynching. Lean bodies, beaten bodies, coal ashes, and the waste of a racist world that squanders human energy. This entropy, this exhaustion, finds its dark crescendo in the cellar of a workman's shop, near the blacksmith's forge, where dead ashes of coal pile on the floor. The signifiers of wasted potential are explicit: spent bottles of alcohol, regrets for acts of debauchery, burned-out ashes, half-burned candles, and misdirected hatreds. Toomer turns our attention to a potential prophet arrived here in the basement, crushed by this misshapen world. He is a mulatto man from a family of orators, struggling to find the right words to seize the world, to wrest it into a proper form, to repair it in the face of a grotesque present. But we find him, down in this cellar, raging to an ancient, mostly dead, black man, used up by the world, sitting in a chair, a man that he calls a "Black Vulcan." The words that come spewing out as prophesy are not enough to change the world—they are "misshapen, split-gut, tortured, twisted"—and they bring him to utter exhaustion, as he collapses, spent, into the arms of a woman, a southern virgin Mary, "ashamed and exhausted."[57]

Climate change is about this entropy—it is about the missed opportunity to burn our fires with justice and compassion and about a racialized heat that was never given a proper form.

1942. A Pacific Island. World War II.

Entropic energy finds its grandest expression in the act of war, in the amassing and exploding of modernity's forces for the purposes of destroying life, for forcing submission.

The world's energy deepening had a decided impact on the scale and nature of each new war in the twentieth century. Jet fuel, napalm, diesel, tanks, bombers, plasticizers, and half-tracks—these fuels and technologies of the fossil economy left a deep imprint on bodies, minds, and ecologies across the globe. Norman Mailer recreates the trauma of this destruction in the shrieking body, the sleepless night, the bayonet, and the hand grenade. But it is war's background, defined by these vast impersonal fossil energies, that disorganize life all around, that wholly disorient the habitus of life. Everywhere we see and hear incendiary bombs popping in the distance, dive bombers buzzing with a wail and a roar ready to unloose their load, and the aftermath of fires burning throughout the jungle. Modernity's energies are in Mailer's text trained on destruction; and they have left in their path wreckage and despair: a demented landscape of humans "burned together . . . black and crippled," "black silhouettes of burnt tanks," and "maroon-colored smoke" rising into the sky after each new assault. Petroleum's waste products, its dissipation and entropy, become for Mailer the fog of war, what he calls "the ornament of battle," seen in the visible residue of burning gasoline gels dumped on the jungle, napalm derived from gelled petroleum, and the brown emissions that follow everywhere the assault craft, the army jeep, and the passing bomber.[58]

What we see in Mailer's depiction of WWII are the propulsive, disembodied, congealed, and polymerized energies of modernity repurposed as agents of death, as thanatos. His novel gives us, in this respect, a glimpse of the irrationality of modernity, of its purposeful acceleration into entropy and waste. Progress, reason, enlightenment do not figure in this insensible military conquest of an insignificant jungle in the Pacific. Mailer's characters are modernity's objects rather than its subjects, people frequently "confused," who "trudge dumbly" onto beaches, who leave behind refuse of "cigarette butts" whenever they stop, and who struggle to stay sane in a natural environment where trees are devastated, "shorn . . . from the shelling," amid the aural terror of "shrapnel cutting through the air, whipping through the foliage of the jungle." Modernity's energies produce panic, reducing us to a frightened soldier "sobbing in [a] hole, terrified," having emptied his bowels in a complete loss of control.[59]

Climate change is facilitated by war; it is policed by war; and it is war's potential. War's entropy is modernity's cyclical failing.

Timeless. Now. The Unconscious.

Modernity's entropy can only be forestalled rather than resolved. There is no perpetual motion. But how we use, manage, and purpose the energy at our disposal is an essential part of what it means to live together and to do so with empathy, thought, and care.

To consider this entropy as we move into the future, we might entertain its relationship to libidinal desires, to the multiplying forces of repressed or blocked desires that are channeled into modernity's forms but never fully contained, satisfied, or represented. Angela Carter imagines for us a scenario in which modernity, this civilization that professes progress and reason, is under siege by the unconscious. She portrays its public space as a totalitarian and highly rationalized one, policed from the city center, surveyed by a Ministry of Determination and its department charged with the Rectification of Names. Hers is a high-energy space of coercive power with weapons of surveillance at its disposal: computer banks, an airstrip base, helicopters, search lights, and teams of motorcade police swarming to root out any assaults on its preferred form of reality. To the extent that this is an allegory of modernity's construct, she portrays it as a formal thing depleted of substance, as something delimited and stifling, where Bach's efficiency is preferred to Mozart's frivolity, where the "thickly, obtusely masculine" has displaced the feminine, and where even prosperity is a "solid, drab," uninspired thing. She offers us up the image of a world of professors laboring, somewhere in the distance, in "steel and concrete villages," spinning their ideational webs to protect that world's form and coherence.[60]

Carter imagines a full-throttle assault on this reified form of modernity. She envisions it coming from the periphery of empire, from the colonial fringes in Africa, from inside the nonnormative spaces of circus tents and peep shows, from trapeze artists, alligator men, bearded ladies, and from a deranged doctor spinning mirages to loosen up the world's pent-up, barricaded energies that have been cordoned off in the novel behind a "vast wall of barbed wire." She imagines modernity losing its form in a surreal landscape where warehouses rise up into "cloud palaces" before fading into their familiar dreary form, where opera houses transform into a babel of giggling and raucous "peacocks," shrieking and fluttering "like distracted rainbows," and where the city's buildings literally become alive, begin to chant mantras, and then change suddenly into "silent flowers." Carter's revolution of the repressed—this great shape shifting—cannot, of course, derive from the old energies, coal, oil, uranium, or methane, controlled by the center. She has to look elsewhere for the power to deconstruct modernity, to what she calls the "eroto-energy" of the unconscious, to the creativity and libidinal desires that have been repressed and untapped—and that lay in wait, ready to be unloosed and transformed into new "concretized desires." To see the world differently, to actualize such change, she thus turns to an unlikely sage, a peepshow entrepreneur, who tells us that one needs to commit to seeing the world differently—it simply takes, he tells us, "persistence of vision."[61]

Confronting climate change requires envisioning and revising, and it depends on the release of new energies, both material and human, to fuel that work. It might not yet be clear where we will get those visions from or where we will find the fuel to drive them. But they will come.

A Coda

Fiction, when reread in light of climate change, breathes differently for us. It opens us up to new possibilities for relating to the past, for reframing the present, and for projecting into an unnamed future. It offers bits of material, natural resources, here and there, new beginnings, new starting points, for constructing a postcarbon self beyond bitumen, beyond petroleum, beyond gas, a self that is ultimately more sustainable and compassionate. But to get there we will first have to retextualize the world.

Epilogue

Carbon's Temporality and
the Structure of Feeling

We live amid catastrophism. A sixth episode of extinction. A jolt from Earth to this new thing called Eaarth. So what is the history of the present? Where did it start? Or, better yet, how do we start it? Where do we see ourselves relative to the past and to an unwritten future?[1]

The historian's business is to mark and flesh out time—to provide us with an orientation to life, a context for living and dying. But this marking, what we call periodization and narrativization, the scripting of beginnings, middles, and endings, is up for grabs. The old comfortable terms—development, growth, and progress—no longer suit us. They are part of a materialist, and often racist and classist, rabbit hole that we have been spinning down (or is it running up?) for 150 years. The notion that we have been building toward something better, healthier, and happier makes little sense, as either history or aspiration, when we look around and see the heavy breathing of the downwardly mobile and the prospect of a sun-baked future toward which we are running blind.

The cultural imperative of development, this historical fetishizing of growth, which over time became simply a synonym for the commodification of life, implies a certain willful ignorance of the world we live in, an ignorance of the ecological and human externalities of such growth. We have been taken, in a sense. Everything we were told about the prospects for our pensions, if we have them, what we were told about laying aside for our children, all of these guideposts we inherit from the wisdom of the Anthropocene, from the great normalizing of rising economies for the white middle class, has turned out to be unfaithful to us. These assumptions, these received wisdoms, were leveraged on a mortgaging of our future, on a temporary feasting on limited resources, ecological imbalances, and the bodies of a people without history. No one was watching the larder (or at least no one with the power to do anything about it), and no one was nursing the injured.

That was left to us.

So here we are—at least those of us who believe in science—needing to rewrite the history of the modern world and looking for a new angel to sort through this debris that has been piling up behind us and, curiously enough, cluttering the road ahead of us.

That work is in our hands.

Marking Time on Eaarth

To rewrite the history of the now first means understanding time and how we use it.

Time is not a uniform thing, as we have been taught. The tick has a different duration from the tock. Critics call this variability in our experience of time *experiential time*, *private time*, or, in some contexts, *temporality*, terms that denote the ways we register time unconsciously and how we reflexively position ourselves in it, either intentionally or unintentionally. We might just as soon claim it as history.

On the one hand, there is the chronometer. Its type of time is regular, punctual, certain—the time of the metronome. Such standardized time is also, at least in one respect, a superficial thing. We know this time intimately. It takes me ten minutes to fill up my tank, forty minutes to drive to work, two seconds to swipe my debit card, and I sit in my office hours for an exact ninety minutes, as I am expected to do. This style of time, which was developed for coordinating business, travel, and history, is measured in standardized units and given the straightforward name *coordinated universal time*. Each day in this system has almost exactly 86,400 seconds in it, except for a periodic leap second used to adjust for slight discrepancies. The world has ridden on this type of time since 1960, or even before that, back to 1884. It closely corresponds to the Greenwich mean, and it is approximated by the clock on the kitchen wall and the lock screen on an Android phone.

But we have not always marked time this way. This type of flat and hollowed-out time has a history itself, and one that is not completely benign. It first gained ascendance as part of what Max Weber termed the *rationalization* of modern life during the ramp up to industrialization. In the United States, we can date its empire to 1883; globally, it dates to the 1884 International Meridian Conference. Those were the years when the nation's railroads established formal time zones across the United States to regulate us to the clock and when shipping interests (with the exception of the French) adopted Greenwich mean time so as to be better able to coordinate longitudinal readings across oceans. This is the homogenizing of time, and it is the precondition of the assembly line.[2]

Yet the chronometer, to be sure, is an artificial imposition, a form of imperial mapmaking that does a certain violence to the world's multiplicity. But it is also a substantial thing that has acquired its own reality following nearly two hundred years of disciplining and accustoming our bodies and travel to it. That is to say, most of the world, whether blue collar, white collar, pink collar, or something else, whether on the assembly line or scrubbing floors or sitting in an office, has learned to watch the clock, to *clock in* literally and figuratively and to gauge our productivity by the big and little hand. This way of parsing time now sits deep in the body.

Yet this measure of time fails to contain our humanity. It comes up short, far short, because it does not comport with how we feel our way through the day and through life. The ten minutes at the gas pump are experienced as an eternity because I am late to a meeting. They feel stretched out, elongated to the point of being unbearable. The forty-minute commute flies by in a minute as I talk garrulously to an old friend whom I'd lost touch with for years. Time is compressed in this other instance—life packed into this small capsule. Children understand this and retain a certain preindustrial preference for such experiential time. My daughter before she was disciplined to the school bell, before violin lessons had introduced her to the metronome, asked precisely the right question when I told her we had ten minutes to get to the park and its swings. What do you *mean*, she said, by ten minutes? Do you mean how long it takes to eat breakfast? To walk to school? To play Candyland? What she really wanted to know is what the next ten minutes would *feel* like.

This other phenomenology of time is a more organic, grassroots, and personal thing. We find it at the root of modernist aesthetics and at odds with the discipline of industry and the chronometer. Philosophers, artists, musicians, novelists have always played with experiential time, stretching it, breaking it, compressing it, as a way to remind us how becoming conscious of time invests us in humanity.[3] For instance, the first English novel, Laurence Sterne's *Tristram Shandy* stretches time to the breaking point. It begins when the main character takes a step down a flight of stairs, and it is only several chapters later, after we have gone down long alleys of memory, reflections on past choices, that we are permitted to complete that step. Such is experiential time. It contains the infinite.

For our purposes, neither of these is perfectly suitable, even if they are both valuable well springs for carrying us through the problem. Between these two types of time, we have what we might call historical time, a form of temporality and temporalizing, that is more intentional. Historical time derives from how we *choose* to narrate the movement of time to ourselves and others—how we become conscious of time, how we adopt certain markers and not others, how we pace out the change over the years, and how we make the past mean something to the present. This is the personal and political war with time. No historian gives equal time to each hour of the day. No person takes stock of the week by giving equal time to each incident that occurred. No history class has ever given as much attention to 1739, the start of the War of Jenkins' Ear, as it has to 1776 or 1789. We select out from the vast repository of the past some of those moments when someone or something speaks above the noise in a way that entices us. And that is a matter of choice, of agency.

How we move from the jumble of life's experiences to shared temporalities, to the assembling of the coordinates of the past so we can mark time together, is, in other words, history's business. It is the act that we designate with a capital H, which is simply another way of saying that History is about telling stories to ourselves, about *finding out*, as it were, an orientation to life, a set of coordinates to live by. So today when the right complains about revisionist history, a term of derision that is increasingly wearying, it is not usually contesting the facts of the case. What it is battling against is the act of marking time for new ends and for empowering new social groups. That is, it understands, as well as anyone else, that History is the work of orienting the self for life, laying the groundwork for agency, power, and empowerment. It is thus political, whether we call it objective, revisionist, radical, or something else.

To be sure, scripting time is not arbitrary. Some holds are barred. The act of making History still relies on the chronometer, because it needs acknowledged reference points. For instance, we can all agree that independence was declared on July 2, 1776, and that the final wording of the declaration was approved on July 4 of that year. There is no need to quibble with this use of the chronometer. It provides some steadiness, some agreed-on anchor for our discussions about the past. But historians also understand that this type of temporality, which is the bane of the history student tasked with memorizing dates, is not really historical time. It is the mere sequencing of data points laid out in a line. It registers the moment before we act, before we meddle in the world, reflect on it, mediate it, and bring it into our experience. It requires that we do some sorting, that we talk to one another, that we sift through the debris and begin to recycle the past for use.

So what temporalities do we have available to us as we re-mark time, as we reorient ourselves, for climate change? What is being offered to us at this point?

The Time of Climate Agency

One way to mark time on Earth is simply to heed the call of climate scientists who in 2015 put us on a short fifteen-year timeline to act. That is the number of years they have given us to retrofit our infrastructure and lifestyle to hold CO_2 below 450 parts per million. If we fall short of that goal, temperatures begin to rise more than two degrees Celsius above preindustrial levels, sending us past a threshold that pretty much everyone agrees will cause a tectonic shock across the world's food systems, fresh water supplies, and ecosystems—all of which rely on the relatively predictable climate of the Holocene.

By the time this book is published, the number of years will have dropped to twelve. By the time you read it, it will have dropped to ten, maybe fewer. Whatever might be the case, we should expect that without quick action toward mitigation, there will be serious changes to the way we do business and relate to one another. The World Bank, for instance, estimates that climate volatility will drive one hundred million additional people into "extreme poverty" and heightened mortality by 2030 as a result of lower crop yields, recurrent drought, excessive heat waves, increased flooding, and rising health risks from malaria, diarrhea, and undernourishment.[4] The World Health Organization anticipates much the same. It plans for an additional 250,000 deaths per year by 2030 from climate-induced malnutrition, malaria, diarrhea, and heat stress should we not act swiftly.[5]

That is, unfortunately, just the beginning of decline. Things get worse, perhaps much worse, from there.

This is a teleology of end times.

The one virtue of this call to action—this marking of time—is that it gauges the present with a certain Platonic clarity. The markers it lays out for us are straightforward scientific ones based on readings of carbon in the atmosphere. Today's climate scientists tell us, for instance, that CO_2 readings hold at just over four hundred parts per million across the globe. The only location that had defied that threshold was the South Pole Observatory, and unfortunately it too recorded in 2016 readings consistently in excess of four hundred parts per million.[6] To understand that gauge is to return to the past. It requires us to resurrect the geological record as a way to anticipate where we are headed. At this point, what we know is that the earth hasn't seen four hundred parts per million of carbon dioxide in the atmosphere for four million years. At that point, we were not yet differentiated as a species: hominins were just climbing down from the tree canopy, toying with life on the savannah, and operating with a four-foot hairy body and a brain the size of a chimp.[7] And the earth was less friendly at that time. It was a volatile place, characterized by frequent hurricanes, cyclones, and, droughts—a place that would throw our hydraulic infrastructure and food systems out of balance. Because climate has a lag time, because the earth responds slowly to changes in the atmosphere, we are already headed into a hot mess, you might say.

There is no reason to quibble with this timeline, but it is not so much history as anticipation. Where is the history of the now that can prepare us for this coming heat?

The Time of Big History

Temporalities arrive swiftly in today's world. One high-profile project for marking a usable past with which to gauge climate change is the Big History Project. This enterprise, created by historian David Christian and backed largely by Bill Gates, offers new coordinates for the present to help us see where we stand in the universe—and how we got here. It aims to bring humanity under the umbrella of one big story that might carry us into today's battle against climate change and against entropy, that is, against what Christian calls "the endless waltz of chaos."[8] To accomplish that, to reorient us to life, it provides us with a single unified account of reality, or, for the romantic among us, one big new creation myth (beginning with the big bang thirteen billion years ago).

Big History takes a swing at life's biggest questions: "Who am I? Where do I belong? What is the totality of which I am a part?"[9] These are important places to start. To answer those questions, it stretches, almost to the breaking point, the scope of history to encompass the evolution of solar systems, multicellular organisms, and the earth's host of sentient and nonsentient beings. In fact, its admirers claim that Big History offers a unified theory of human knowledge, the whole kit and caboodle, akin to Darwin's theory of evolution.[10] As Christian explains, the project itself strikes out for a "workable map of reality" to tackle today's biggest problems and questions with humans rather than tribes in mind.[11]

So let's entertain this notion and go big temporarily.

Big History maps out a few main coordinates to steer us through the world. Five of those coordinates stand out above the rest.

First, Big History places humanity's origins in the big bang, in the cold void of a *something* emerging out of nothing. It begins, perhaps not surprisingly, with the formation of the universe, 13 billion years ago out of emptiness, through its inflation, its spinning globules of hydrogen and helium, and its formation of an asteroid-dented earth 4.6 billion years in the past.

Second, Big History takes us to the formation of life on earth, to a deep thermal vent somewhere at the bottom of the ocean where, not long after the earth was created, archaebacteria quite likely crossed the threshold of life for the first time.[12] It sees this magic of chemical animation, this shock into life, setting in motion the development of early multicellular organisms and the later complexity of high-order organisms, a world of birds, tigers, camels, and apes.

Third, Big History takes us to the moment when *Homo sapiens* began to rise above, to break off from, the pack, to 250,000 years before the present, when genetic mutations provided us with attributes like symbolic language and expanded social networking that allowed us to outpace and displace many biological competitors, including Neanderthals and other hominin.

Fourth, Big History jumps forward to the birth of agriculture, in the warmth of the Holocene, to 11,500 BP when our strain of humanity experimented with sedentary farming, this halting development that led over time to surpluses, cities, and rising reproduction rates. It sees that development as initiating the division of labor, hierarchy, and social complexity that defines the world today.

And, finally, Big History carries us into the Industrial Revolution, into the emergence of modernity 250 years ago, when we sprang into a high-energy world characterized by wild degrees of wealth, congestion, and decay. It locates the we (the who we are today) in this spike in the world's population and economy (anchored in fossil fuels, science, and capitalism) that laid waste to preindustrial hierarchies, parochial cultures, and traditional political forms.

Here it lands us in the blessing and the curses of the now.

To mark time, in this way or any other, is to have a project, and the Big History Project is no different in this respect. After all, choosing to skip 3 million years here, to settle on a moment 250 years there, is an act of intervention. Part of that project is straightforward and healthy. The Big History Project retrains us to see the forest for the trees, to look up from the glove box, to climb out of the archives, to rethink which details in life, in the past, are important so that we might see what we've missed because we weren't looking. That is, it puts our specializations in their proper place. It does that both by reuniting the humanities and sciences, which have for too long been sitting on opposite sides of the stadium, looking across an empty field with no agreement on the lines, the rules, or the meaning of the game, and by transcending the extreme division of intellectual labor that has over the years eroded portions of our humanity.

But Big History's creation myth comes with obstacles. It is, for instance, a top-down history, imposed on the world from above. It reinforces a certain passivity in the learner, training us to accept History as the memorization of a new canon of dates, a new master narrative, albeit a disruptive one on a much longer timeline. The trouble is that History, at its best, has always been about asking questions, about assessing evidence, about analyzing documents, and thus about cultivating critical consciousness. On these counts, Big History falls short, inculcating instead a hands-off approach that cedes our memory to the experts. The result is an architecturally elegant but abstracted venture that is, like all other modernist achievements, whether Brasilia or the Green Revolution, out of line with how people see, feel, and labor their way through the world. It is Kantian rather than Hegelian, and thus for all its brilliance, the Big History Project is unlikely to elicit the empathy that might mobilize us to action. By being detached from our emotions, from the life of our tribes, it loses something vital, and, to be sure, if we have learned anything in the last few presidential cycles, it is that people don't get moved by abstract logic: they need to feel it on the ground, they need their knowledge to touch their affective communal structures.[13]

Good history is read and needed because it lands on something and in doing so mobilizes action.

The Big History Project leaves us wondering how we might derive incentive from single-celled organisms growing near a deep-sea vent or how we might wrap our minds around a universe that doesn't care for humans or even find them to be a priority.

The Time of the Anthropocene

A second stab at the problem, at marking time for climate change, comes from those writers who tell the history of the Anthropocene. This alternative mode of writing the history of the now narrows the gauge down to the present by telescoping in on us as we have unfolded in the last century.

This way of marking time also has clear origins. The term traces back to a comment made by the atmospheric chemist Paul Crutzen at an academic conference on climate change in 2000, in which he contended or, more accurately, blurted out, against current professional opinion, that we had departed the Holocene and entered a new epoch defined not by geological forces but by man-made impacts on the planet. From that point forward, the term picked up a life of its own, entering into serious geological debates, policy debates, and even literary criticism. And although there persist legitimate differences of opinion as to whether or not the Anthropocene began back in the Industrial Revolution when carbon dioxide emissions first began to heat up the planet or more recently in what historians J. R. McNeill and Peter Engelke term the "great acceleration" of the post–WWII era, the terminology shapes institutional thinking about the present—about its meaning, about its boundaries—in both the sciences and the humanities.[14]

The Anthropocene whittles down the origins of the present, dating it to a much more recent time. If we accept McNeill's and Engelke's definition, it begins in the hyped-up economy of World War II, whose aftermath led to a drastic population spike and a corresponding spike in wealth. The argument is that human industrial practices and their corresponding consumer rites pressed such a deep stamp on the earth in this period (registering in increased ocean acidity, global climate volatility, and planetary deforestation) that, like a layer of iridium left behind by an asteroid, their effects will remain visible on the planet's geological stratum for the rest of Earth's history or at the very least will "linger for millennia to come" in its biosphere, atmosphere, and lithosphere."[15] The Anthropocene refers back, in other words, to a geological, or, in some cases, ecological, gauge for marking time.

There is little doubt that the Anthropocene strikes the right mark for our geological time.

But as a way of writing history, the Anthropocene finds itself hamstrung by many of the same problems as Big History. It too is, for instance, a top-down analysis aimed at earth systems as much as at human affairs. That makes it chiefly a diagnostic tool, a record of humanity's material impacts on the planet, one that touches down only here and there on our shared history in the other ways in which that history means something to us. McNeill himself is attuned to this slippage. He clarifies for us that we need to make an analytical distinction between writing the history of the planet (i.e., the Anthropocene) and writing a history of people (i.e., the multiple histories currently on the shelves). On the one hand, he says, there can be little doubt that Earth itself is passing through a new stage of its history. But it is not clear, on the other hand, he says, that human history (that is, the ways we experience and mark time together) is going to correlate, except loosely, with this geological and ecological event. McNeill says that he expects in the near future that the erosion of modernity's ecological basis will force us to define the self around the Anthropocene, but, for now, he leaves that an open question.[16]

The Anthropocene might very well become our history. Whether or not that will happen, however, largely depends on how we weather the storm ahead. If it turns out, as McNeill and others suspect it will, that ecological pressures come to define who we are as a people, then the Anthropocene will be the framework for understanding the meaning of our lives. On the other hand, if we manage to avoid collapse, if we avoid widespread system failure, then we will instead make history inside of the Anthropocene without having to feel its pressures and to name those material pressures day in and day out. There is a chance, that is, that we can mitigate, decelerate, cooperate, and discover a human trajectory driven by something other than reaction to crisis.

That is, after all, the hope. Isn't it?

The Time of the Capitalocene

A third (and for us, final) temporality of the present goes by the name of the Capitalocene. Like Big History and the Anthropocene, this metanarrative stakes out the historical terrain of the now to position us in a new lineage of time adapted to crisis and its mitigation. As the name suggests, the Capitalocene lays the problem of climate change squarely at the feet of capital and its corresponding regimes of value.

This concept of the Capitalocene, both new and familiar, originates in the writings of sociologist Jason Moore and the emerging leftist critique of the Anthropocene. The term itself, Moore explains, is part of a twenty-first century project to "name the system" and, more precisely, to critique the Anthropocene's inattention to capital and its metabolism.[17] While the timelines are fuzzy and the history at times impressionistic, the Capitalocene locates the beginnings of today's climate crisis not in the recent industrial past but somewhere deep in the long sixteenth century (1451–1648)—in a time well before the shift to fossil fuels, during the West's break with the medieval world.

According to Moore, the Capitalocene traces its origins back to two genetic mutations that accelerated capital's speciation into the climate-altering beast it was to become. First, he explains was its emergence around 1450 in Europe as an expansionary project for organizing both human and non-human nature. The agricultural revolution of England and the Low Countries, the colonial sugar-slave projects of the Iberians, the rise of the grain and timber trade in the Baltics, the expansion of mining, metallurgy, and deforestation in Central Europe introduced, he says, a new scale, speed, and scope to the transformation of nature under capital's auspices that was nothing less than epochal. Moore suggests that these developments, in fact, had a greater impact on history "in relational terms" than did the rise of the steam engine. Second, Moore posits a parallel epistemic, or even ontological, mutation taking shape in this period that saw capital absorb into its project the Cartesian principle of dualism from the Scientific Revolution. That principle, he says, was also pivotal. It created a framework for legitimizing the exploitation of the environment by ejecting nature from society and shrinking it down to something external, mappable, and irredeemably other. It was this dangerous combination of local developments in Europe and their subsequent globalization, he contends, that created the operational and ontological basis of today's environmental crisis.[18]

As a form of critique, the Capitalocene has some advantages over these other narratives. Most importantly, it highlights historical processes by returning our attention to the fact that climate change was not a preordained consequence of industrialization nor the result of irrepressible Malthusian dynamics but instead an act of collective class agency, the result of a certain ideology and praxis for organizing the world we live in. Capital might have trained us to think that the exhaustion of the world is a perfectly natural outcome of socioeconomic pressures (and we might have swallowed that fish whole), but the fact is that nature and society—and their limits—are, as Moore tells us, mutually constructed within a "web of life" that does not distinguish between the two. The trouble with life under capital, we learn, is that it has perpetuated for five hundred years a faulty framework for seeing and talking about nature and society, a framework that disavows the kinship between the two, while its internal logic (i.e., its mechanism for overcoming the law of underproduction) has pushed us, time and time again, into developing new unsustainable frontiers of "cheap nature" located on the edges of its system, where nonmonetized resources and noncommoditized labor can be mapped, exploited, and appropriated as a formula for continuing surplus accumulation.[19] Depletion, exhaustion, and crisis are, in other words, not normal. They are the by-product of a specific class-based formula for being in nature.

This is possibly good politics. It is certainly good theory. But the Capitalocene also presents us with a couple of problems for making time in the age of climate change.

There is, for instance, an academic aspect to the problem. The Capitalocene doesn't fully jibe with our history. We can't, with an eye to the full historical record, iron out capital's uneven and erratic past by positing an origin, as early as 1450 AD, when you and I were set into motion. Capital's project only looks in hindsight to be more accomplished than it was as it unfolded in real time. Absent the thermodynamic revolution we really don't know where capital would have taken us. History and theory part ways here. The former, which insists on contingency, tells us that capital developed through a type of punctuated equilibrium whereby an initially small and unanticipated historical mutation (i.e., the fossil revolution) accelerated its speciation into something radically different, something even more dangerous, than what it had been up until that time. Capital exhausted nature on a planetary scale and created global climate change not simply because of its own internal logic but because that logic was propelled forward by an unthought, and even unthinkable, subterranean frontier of work/energy, a resource windfall, that remade its metabolic infrastructure—and consequently grew it to size. Not all frontiers are equal, and the fossil frontier, which lifted the prehistoric past into human history, was something quite different in measure and kind than simply opening new soil or integrating new peasants into the system.[20] A second, nonacademic aspect of the problem is political. The Capitalocene leaves only a little room for historical agency. In its accounting, the names, and the people themselves, disappear from the historical record in favor of theoretical abstraction, while the past and future shrink down to a familiar and seemingly inexorable Marxist history that wraps us all into one neat package and that replicates modernity's own troubled tendency toward high architecture, a type of totalizing that might even rob us a little of our humanity. We find ourselves stuck, in a sense, waiting for and anticipating depletion and crisis, for the well to dry up, for the fossil frontier to exhaust itself, before the revolution comes.

To be sure, the Capitalocene holds out some political promise. But it tends toward a history without author, a process without people, and a story of the now that might be as debilitating as the ones it critiques.

Making Time for Ourselves

The historical models we have for marking time today are, to some extent, emotionally inert, maybe a little bit tone-deaf culturally, and certainly detached from the pressing time frames of life and the more immediate manner by which we live through time. To be clear, this is important work. Climate change needs these metanarratives to rally policy makers to the task at hand and to speak to universal problems. But to become real to us, climate change also needs to be liberated from such monolithic, and preordained, timelines—from such abstracted causality. It requires a degree of creativity in our history making that fits within what Raymond Williams called the "structure of feeling," or, in postcolonial terms, that fits within new structures of feeling.

So how do we make time for ourselves? How do we find the right origins, mark the middle, and project forward?

To locate a history that touches on our aching and aspiring—that lands on the heart as well as the head—is to migrate backward from the planetary to some version of the local. It is to reintroduce into our story more nuanced questions of identity, of nation, class, race, and gender, and to recuperate ancestral memories of resistance and conquest, of sustainabilities and of failures. The present will not come to us in neat packages, and it will not be of a piece; it will be written by different actors to empower different ends. It will be defined by temporary constellations of time around which we can collectively reorient our lives before moving on to the next. The size of that history is negotiable. There is no preestablished canon of dates, and we have no Kronos to set the clock for us.

So to conclude, let me offer something less sober, less totalizing, and more open ended. But, better yet, each of us might play with the following question. What are the events—five, six, ten—around which we can rewrite the history of the now, by which we can reorient ourselves for life in the coming heat? How can you and I mark time together?

The Ice: 13,300 BP

America begins in Alberta, not far from Glacier National Park. There are no territorial lines. No state. No wilderness as we understand the term. Most of this time is lost to us. But we can catch glimpses of our origins, here and there. One comes to us with rare clarity.

A tribe of us, Asian American hunters, is seen on the tundra butchering a camel, eager for its eighteen hundred pounds of meat and bone. That camel, an ice-age progenitor of the arctic camel that once thrived in America under warmer conditions, will soon disappear from the continent and it won't return for another twelve thousand years. We have, as a tribe, already taken down at least seven other horses.[21] And we have been traveling from somewhere, no one knows exactly where, but across the north, leaving footprints in the snow and a trail of carcasses. A wooly mammoth. A giant sloth.

The ice is still thick. Glaciers sit deep in what will be the Great Lakes. Home is far away. Or perhaps it is right here.

At this point, the Laurentide ice sheet has not yet broken. The earth is tilting and we know that it will reach a breaking point—but not yet. When it does, when the warming comes, Lake Agassiz will flood the St. Lawrence, and a vast glacial retreat will leave behind the Great Lakes. For the time being, for now, we have not strayed far south from a place to be called Calgary, and we hunt without flint, without Clovis spears. Eventually, however, our desire for protein, berries, nuts, takes us gradually southward, where we will lose the route back to where our ancestors started, and where we will lose the desire to return anyway.

America begins, that is, back in Asia. We are all Asian, or perhaps all simply human. But we begin together, hunters and gatherers all, chasing after protein and fire, leaving a path of extinction in our wake. In the near future, megafauna, these now-extinct mammals, will fall and won't return. The horse will be gone until its distant ancestor is brought back by Europeans. When the American camel returns, it will be to the zoo.

This is where our first peoples begin—in ice and snow, in the hunt, in extinctions, capitalizing on a deep freeze.

The Warming: 10,000 BP

Millennia have passed. We turn south, and look up at the Andes. Or perhaps it is down into the Tehuacán Valley of Mexico.[22]

Someone has kicked free a patch of loose dirt and dropped a potato bud into the soil of the foothills. Someone else has dug a hopeful kernel of corn into a fertile plain in the north. These are the seeds and buds of empire. They whisper of surplus, hierarchy, of power. The potato might have been purple; the corn, when it is ripe, will be the size of a human finger. For now these are tentative, even desperate, things, but they will in time blossom into the world's first and fifth largest crops—become confident, arrogant— the basis of the division of labor, giving rise to poets, artisans, slaves, surplus, and stratification.

Our sowing comes late—after the ice in the north has melted. After the Great Lakes, the St. Lawrence, the Mississippi have become fixtures on the continent. We farm with reluctance and nomadic tribes still roam the map. But the climate has a warmth to it now, and some of us have entered into a new contract with the soil—one of digging, waiting, and praying. The sun, the moon, the seasons—a trinity of maize, beans, squash. We operate with no instruction books, no roadmaps, no precedent, and no sense of where all of this tilling and storing will lead. Yet from this little trial in the warmth of the Holocene, history's direction starts to take root, with premonitions of beauty and ugliness, hope and entrapment.

From a stockpile of corn, empire grows, settles into itself, claims a periphery. Tenochtitlan. Cuzco. A million of us. From planting and tribute, a new dense world of sociality and materiality, of aqueducts, elevated walkways, floating gardens, fertilizers, sacrifices, and subjugation emerges—a world that operates without wheels, without engines, but with a new degree of satisfaction and complexity, with pictorial writing, art, religion, architecture, domesticates, deep hierarchies of caste, and new political structures to subjugate.

Empire entices and coerces—it subjectifies—the willing and unwilling. But everywhere this new contract with the earth, this new arranging of one another in stratified networks, thrives on labor and bodies—on a thickening concourse of sociality and discipline, of thin bodies, beautiful bodies, grotesque bodies, fattened bodies, tall bodies, and short bodies and on feasts of tortillas, beans, chocolate, and chilis and famines that deliver some of us into slavery. These imperial pretensions will stick. They won't go away.

Not far off is the Virgin of Guadalupe.

The Conquest: August 13, 1521 AD

We leap forward. We prefer to live on the fringes. But increasingly, we are not able to make that choice. The option is eroding. Time, like society, is being compacted, the trees becoming hard to distinguish from the forest. Powers come and go extracting surplus in the name of princes, artisans, and poets. There are, however, big surprises in store. Sometime in the month of July, in the year 1519, marked by the Julian calendar, the world cleaved in half, at least that is what it felt like to us. Perhaps it had simply come back together again.

Cortés's arrival has never been satisfactorily put to rest.

Avaricious. Desperate. Human. What we know is that his disembarkation on the shores of the Yucatan, after a mutiny in Cuba, set in motion a confusing jumble of invasions, alliances, robberies, slaughters, and enslavements. Love. War. Rape. In hindsight, one day seems portentous. It sticks in the craw. Our envoys had met them for what was a friendly cultural exchange. The captain and his men took off their helmets. They put down their swords, left behind their snarling war dogs and horses. They presented us with a heavy metal object, a cannon, or *cañón*, by its Spanish name. We looked on as their soldiers of God stuffed *el cañón* with fire and exploded it, as if by magic. The Codex tells us that many meters away a tree was "pulverized." Moments later a mountain face was "dissolved" by the blast. Why, we thought, would they blow up a tree? Why a mountain? What was this *gratuitous assault* on nature?[23]

We could not have known it at the time but that performance was meant for us. It was a warning, an augur of submission.

The origins of our planetary crisis start here. That crisis begins in imperial conquest—in a many-centuries-long ritual of naked violence under-written by ignorance, selfishness, and collaboration. What followed was only partially intended, set in motion by forces outside of ourselves, by pure accident. Smallpox, malaria, yellow fever. Still, much of it was willed for the greater good—tomatoes, sugar, and rice. But it is the part hurled forward by choice and by force that needs the most untangling, unknotting.

Colonization is the pretext to climate change and empire is its milieu.

An Awakening: 1705

History is picking up speed. We are left in its wake, but sometimes we pause to reflect, to take stock of ourselves. So let us return to a sleepy inci-dent on a farm along the Hudson River Valley in 1705.

In the spring of that year, one of us, a Dutch tenant farmer, discovered, on an eroded hillside, a five-pound petrified molar. That tooth sticking out of the soil was the size of a fist, too large to fit into any preconceptions we had about the world. What went through his head, what wheels turned, at the discovery we can't know. What we do know is that this tooth was later traded for a glass of rum and sent on its wending way through history, land-ing in the governor's mansion and circulating across the Atlantic where its inscrutable origins became the topic of serious scientific discussion.[24]

That mysterious tooth from the great American incognitum materialized out of an indigenous dream from long ago. Uncanny. It caused us to wonder the unthought, about what kind of unknown world preceded us. What world, we were compelled to ask, could have known these teeth and these great rib bones found subsequently that looked "like limbs of trees." Not immedi-ately, but perhaps soon enough, we pieced together a puzzle first laid down in the Ice Age. We rediscovered a part of ourselves we once knew. One of us, the American naturalist, Thomas Jefferson guessed that it came from a "mammoth" or a "great buffalo," six times the size of an elephant, probably living out in the great American desert. A friend of his at the time, George Washington, picked up a similar tooth, a "grinder," he called it, from the Big Bone Lick area of Kentucky and displayed it in the entryway to his Mount Vernon as a curiosity, as an object of contemplation.[25] That is to say, in the midst of war and revolution, some of us were wondering about something bigger.

The answer came as a thunderbolt from across the Atlantic. Word ar-

rived that we were looking at palimpsests of God's creatures lost to time—
that these teeth, rib bones, and skulls derived not from living elephants or
any other still-breathing creatures but from extinct souls, beasts, mammoths,
no longer accessible to us, gone, vanished from America. Ergo, the rediscov-
ery of life's ultimate precarity, the birth of paleontology, and the resurrec-
tion of extinction. We saw a sliver of life outside of ourselves—we came to
understand that things die out, that they don't come back, and that the
earth is not cyclical but rather catastrophic—that it has a direction and that
we play a part in its direction.

Here was the incipient *a-ha*, the exclamation that God had left us to our
own devices, that humans could coax the world to extinction, and that our
ancestors were not who we thought they were. We had discovered the beginnings of the modern soul.

A Gain and a Loss: 1776

We see that modern soul, however, through a broken mirror, in fragments, in bit pieces of DNA. Still, some fragments, some genes, are more
assertive, more pronounced than others; they dominate the picture, impress
themselves on us, bully aside the rest. That is the case of a unique mutation
that redefined the soul's instincts, taught it to measure out life, spoonful by
spoonful, through the narrow strictures of the market, to remold itself to
this die-cast *Homo economicus*.

The year 1776 has some claims to us on this score. For in that year economic man found its voice, its articulation. We took on a new raison d'être
in the midst of the ferment of colonial revolution. Amid patriots clamoring
for freedom in the colonies, with empire tarred and feathered, a Scottish
economist, across the pond, summoned us to a new reason for life, to a new
and exciting game plan for stripping off the straitjacket of empire, for shed-
ding the constraints of closed colonial markets, and for opening up a new
era in human creativity. We heard what we wanted to hear—that our getting and spending, our pursuits of individual happiness, had a moral purpose to them, that they could be scaled up without consequence, that we did
not need God to drive the world, that other invisible hands could be trusted.[26]
He challenged us to dream big, happily, and maybe a little bit sloppily, and
those dreams landed on receptive ears in the excitation of revolution.

Sometimes the best of intentions fail us.

This evolution in the soul came late, and it comported with something already there in us. It felt natural to us, even if it marked a moment of punctuated disequilibrium. After all, we had always traded, always capitalized, so to speak. The market had been a place of joy and carnival. Local eggs, handspun wool, deer hides, alewives, and acres of land deeds. It had also been a place of suffering. Of bodies for sale, families torn asunder. Small, local, and personal here. Big, global, and abstracted there. But until now, the market had always had a time and place for convening and adjourning. Its mentality was marginal to, something set aside from, the rest of our lives. If a few of us, a small strata of merchants, bankers, and slaveholders, lived and breathed by it, oriented their sense of self around it, the fact is that the rest of us—yeoman, squatters, slaves, small-town dealers—kept only a small hand in the business of doing business. And yet we were coming to see, to even believe in the heart and head, that the act of privatizing property, cordoning off things, accumulating, and getting the upper hand on someone else could be a way of life—a better way of life and a source of morality.

Along the way, some of us looked down on what we had made and said it was good. Very good.

A Deepening: 1846

But the genealogy of the modern soul is more peculiar than that. Capital is one thing, industrial capital, or to be more precise, fossil capital, quite another. We have often been confused about the two.

The modern soul has its origins, it meets the crook of the trunk, at a particular place and time—in the shift away from renewables, from a type of growth easier to comprehend, to keep properly pruned. To be sure, the fossilized branch in our genealogy started out humbly enough in some cellular growth that revealed no real portent of its future overgrowth, but soon it would crowd out the tree itself.

We have known carbon for a long time. Humans have been burning it from the time Prometheus stole fire from the gods. Fossilized carbon—coal, oil, gas—arrived, when it arrived, as a mere extension of that past, a replacement for wood, nothing new. When trees failed, coal sustained London, Boston—an ersatz fuel for a hearth that preferred to burn sweeter stuff. None of us looked, at the time, on coal as the fulcrum of freedom, of labor power, of demographic growth, and none of us knew, could have known, this shift would steer industry and economy into the seeming infinite, beyond the familiar ecology we knew—into a future piled with riches, driven by human overreach of its habitat.

We can, in fact, be quite precise about this. While the fossil lineage traces back to many points of origin—Watt's steam engine, Slater's mills—these familiar markings are a misleading tangle of the organic and fossil branches in our genealogy. The modern branch of the soul thickened in North America at a particular time and in a particular place: at the Massasoit Mill in Fall River, Massachusetts, in 1846, a mill curiously named after our displaced Indian fathers. Remembrances of things past. That was the year that we turned the corner, put together the pieces of the industrial puzzle. The Quequechan River tapped out, lawsuits over water rights, a shift from hydropower to coal from the mountains, to steam power's predictability. We would reach 1.2 million spindles, outpace Lowell. Industrialism, this churning of wealth, no longer looked quaint. The dream of an "industrial pastoral" along the Merrimack—a more humble mill world of hydropower, dug canals, recruited mill girls, new boardinghouses, and promises of temporary wage labor—got shoved aside for something bigger, newer, more ambitious and more permanent.

From little seeds things grow.

In the years to come, we would write it out in frank letters: "In Steam We Trust." We would shelf the waterwheel, leave behind the stream, forget restraint, and offer ourselves up to the Tartarus of Maids, dreaming of a Paradise of Bachelors along the way.

Pruning the Tree: 1863

But history is not only done to us. We sit astride it. We ride it. Sometimes with great control, with command. More often like a bucking bronco—hanging on for dear life, maybe even regretting the ride.

There was always this question (at least in hindsight) as to what fossil capital would look like once the chokehold on the energy supply was taken off, who we would become once we left behind the topsoil for the subterranean and embraced this new eco-logic and its economy.

A proposal came from Gettysburg, amid war, with bodies lying limp in the fields.

The big question concerned the rules for the modern soul. What principles it would live by to mark out a meaningful and righteous life. Would we find form within a sleepy violent world driven by a racial caste system, class propriety, plantations, and whips or within a raucous industrial one characterized by deep pits of carbon, hyped-up mechanization, wage discipline, and rising surpluses? There were other options. But they were not considered.[27]

It took six hundred thousand lives for us to decide by point of bayonet on fossil capital—on wage labor, fuel-driven automation, and industrial management—as our preferred form of becoming. We would revoke that decision here and there, and we would struggle to acknowledge the persistence of inequalities of class, gender, and race, but we would at least commit ourselves in principle, in 1864, to a certain type of equality and a certain type of freedom outlined by our spokesman, tall and lean. Let us call it bourgeois freedom, liberal freedom, laissez-faire freedom. Whatever the case, fossil capital would offer up its own rules for emancipation.[28]

Light had in the past radiated from steel chains, but here was at least a proposal for something different. The modern soul would take shape inside a new imaginary, a new, often invisible *iron cage* of oil wells, coal pits, cities, factories, and railways, of intense accumulation, of industrial discipline, and equally intense austerity. We would expend ourselves on that dream in faith and fury, but it would burn bright in America for more than 150 years.

Industrialism by coal was the basis of a dream unfulfilled, and today we are left to ponder whether it will dry up like a raisin in the sun or sag like a heavy weight upon us.

A Recognition: 1938

The modern soul is expansive. It glories in the infinite. But it too must know that all things bright and beautiful have boundaries, that they require sustenance from some soil.

Fossil capital collapses (and it panics, it breathes agitated and quick) with regularity. 1847. 1873. 1893. 1901. 1907. 1929. Panic and collapse have typically been a result of the failure of human agency. Collapse generated by us, and for us, and for reasons we can understand: bad decisions, a lack of empathy, insufficient understanding of the market's machinations, speculative bubbles. This recurrent collapsing of fossil capital might be acute, terrible, harming, but we can hope that in managing it our agency matters. We can even dream of it collapsing so as to give way to something fuller.

But *ecological collapse* is quite another thing. A form of terror that makes the whole earth and body shake. Fossil capital, with a ferocity unlike its premodern counterparts, can today summon forth a collapse of more catastrophic planetary proportions. We have already known it locally.

The rains died out on the Plains in 1934, after many of us had ventured onto them, envisioning markets and boom times, during a wet period of swaying grasses. We pushed crops beyond need. Soil beyond comfort. Markets beyond saturation. Our migrations brought with us, carried in our portmanteau, the belief that good soil meant good living, that rain follows the plow, that hard work, sweat, means prosperity—and that fossil-fired railways, tractors, harvesters, and slaughterhouses would knit the plains, its crops, its grasses, its cattle, into a national and transnational marketplace of opportunity. But dreaming has a material context. We had forgotten what we once knew—that west of the hundredth meridian was a place of little rain, a meager place of long horizons, once called the Great American Desert. Here trees dwarfed, rivers ran far and few between, and rain refused to make any promises. Yet we found ourselves here all the same, with our hungry families in tow, having rewritten the name of this place on the map in hopeful aspiration—redesignated the semidesert as farmland, home.[29]

The Dust Bowl came as a warning. Fossil agriculture had torn up hardy, drought-resistant grasses, extended deep into marginal lands, replaced nature's resilience with fragile and fickle crops. The Plain's evolutionary beasts, bison, had at the same time been displaced by these more flaccid and tender things, cow and pig, beef and pork. We should have known, but we didn't, that it was only a matter of time before the well dried up and pain rained down on the plains. Dust masks for fragile lungs. Banks of sand. Tractors stranded. Environmental refugees on the road. Little buds of corn straggling through what was left of the soil. A great wasting away. Cattle slaughtered and rabbits clubbed to death in a great human venting, in a collective anger at nature's audacity, at our own limits.

Fossil capital delivered us into a new desperation. On the American plains, entropy was revolutionized in eager expansion, followed by deep inward withdrawing. Resilience and then abandonment. Route 66 to a life of orange trees.

We might remember a time when collapse meant migration, when failure meant searching out new lands, finding subsidies, a loaf of bread, a cornfield, from elsewhere. Today there is nowhere to go, the ground has been hoed full to the fence. We are all in, and we will have to make do with what is in front of us, whether it rains or not.

An Existential Pause: 1945

To be modern in the way that fossil capital urges us to be is to have hope in growth, in the human capacity to reach toward betterment, to embrace progress, even in the face of collapse. But there are times when we wonder, and we have done it together out loud, whether amid this teeming world of Priuses, deep refrigeration, and episodes of *The Tonight Show*, something might be spinning awry, off kilter.

On August 6, 1945, we gasped at what we had done. Three days later we sighed.

There is, of course, much in modern life to take pride in. Things to feel good about. We have, with great success, remade the conditions under which we can thrive, rewritten the script between the soil and the self—quite literally invented a new ecology to live on, one that has a richer, more vertical, and denser potential. We have galloped together on a little dust cloud of carbon. And the trail has been something to see.

Yet to know energy—to release it—is to know power. Tapping deep energy, reanimating fossilized life, had nonlinear consequences that we are still sorting through. In unloosing the oil well and the coal mine, we cracked open a type of existential energy that brought us face-to-face with the incomprehensible. Today's dark matter appears to us in the unscriptable chaos of climate change—as the slow, partially legible violence of eroding resources and the diminishing hope of modernity's aspirants faced with this newly weirded, and estranged, Eaarth. But we have seen these premonitions of death and power misconstrued before.

It is all the same story.

Fifteen kilotons of TNT. Energy of insidious intent. It was called affectionately "little boy" when it fell on Hiroshima, but we know that little boys are more innocent than this. Watches stopped a mile and a half from the epicenter, on the same small hand, at the same second. 8:15 a.m. A blast that literally stopped time.

In the aftermath, we walked the streets to see human shadows pressed into blanched cement. Flesh protecting concrete, in a strange inversion of life. Eerie ghosts of decisions made. We walked through a city disfigured: 356

years of accretion returned to molten clay, piled in shattered cement. Even more difficult to process was the slow internal corrosion that followed us like a silent predator. Fear in the very atoms of the flesh. Invisible leakage. Hair falling out. Mutation. Marriages canceled. Unknown changes in the DNA.

Seventy thousand of us gone in a flash . . . and as many more in slow death.

Today we anticipate a world where extinction is not only possible but probable. But we once heard its premonition in this great planetary yelp.

The Millennium: Now

For now, for the time being, for the short future, the modern soul radiates. Sun still shines on a red wheelbarrow. It is a sight to behold. Seven billion of us, 1.3 trillion tons of corn a year.[30] A skyscraper, made of sheet glass, that rises 2,717 feet out of the desert. A temple to Ozymandias. Yet it is possible for dreams to die in a blaze of glory as easily as it is for them to perish in ice. Today the soil is hot, baked. It worries. And it is, maybe, a little tired.

Nostradamus warns us that some of us will not be alive at sunset. Madam Sosostris cautions us to beware the unreal city, warns us of death by water. Blind Tiresias, old man now, calls us back to the warnings of the dead, the voice of the bird. And today's prophets—the collective we have put our trust in—pronounce that oranges and avocados have something to fear, that wheat fields will have to migrate, and that we might already be environmental refugees with no place to go.

We near the present. In 2001, at a particular time and place, at multiple times and places, word came from everywhere—from Australia, Morocco, Sweden, Italy, Germany, Brazil, New Zealand, Canada, Ireland—that the relative stability of the Holocene, this period that gave rise to writing, agriculture, markets, empires, and steam engines, had seen its end, and that we, together, had abandoned providence for a world of our own making.[31]

Today, we have taken the soil and the air into our hands. But it is not as easy as that. Earth, Gaia, Pachamama, Erde insists against being directed, on being demoted to a support role in the drama. It issues forth a great refusal to sink into the background, to simply fill out a scene or two. The dry west is slated to get hotter, with longer droughts, more frequent floods, and quicker snowmelt. The soggy east is expected, against our wishes, to get wetter by 20%, to show its teeth more frequently.[32] The seas will rise, and we don't know how high. The summers are already getting longer and more acute, the winters unpredictable.

In such a revised context, we will be left to ask millions of small and big questions that once had answers. Some will be personal. Can I plant a peach tree here, on this slope, at this time? Can I expect the required frost days that it needs to hibernate? Will spring announce itself with enough clarity to signal it is time for budding? Can I build here, in this branch of stone pine, a swing? Does this once-confident pine belong here in this new climate of Eaarth? Or is this now a place for dwarf trees? Has it become too brittle from years without rain from infestations? Other big questions will follow on the heels of these small personal ones. Can the city sustain itself on unknowable snowmelt five hundred miles away when rains come with less frequency and when snowpack thaws under a more aggressive sun? Will the cement of this inner city be hotter, maybe too hot, next year, or the year after that? Will some of us need to plan for coastal retreat from the cliffs, from the island? To move nuclear facilities away from the ocean?

Will it be time to migrate?

The readings of the cards come in quick succession. The Drowned Phoenician appears. Nearby, the signs are already apparent. Two blocks away, the street pines have given way to the bark beetle. A microclimate of shade lapsed to the heat. The zebra mosquito breeds in drops of water here for the first time in memory. Zika begins. From the city center to the rich suburbs, the 200-year-old California dream of green lawns in the semi-desert, visions of David Hockney, is being torn up, night by night, lot by lot, body by body, for a bare landscape of disaggregated granite and spiny yucca. We have our sleeves rolled up and are preparing for a return to the desert. Not far off the nation's Central Valley is baking dry in a matter of life and death. And for a short spell, we even took, together, a collective breath when, in an unthinkable moment, before the rains finally fell, we all wondered if the world's sequoias might die out, if 3200 years of persistence were all they had left in them, if we would see collapse, the beginning of extinction, in the naked eye. Death by fire.

The earth, like a stream, is never the same thing twice: nor is the human. We live through beauty's impermanence. That is its nature, our nature. We are not a prophesy set in stone; and humans have already been many things in the past—hominini, Neanderthal—and always something has survived, a little part of ourselves passed forward into these new constellations of the collective self.

When the world comes at last to fire and ice, there will have only been one of us. The modern soul will have only been identical to itself. But to lose that soul, to watch it expire, is not the same as losing ourselves. It is nothing more than the opportunity to re-baptize ourselves in new waters, to re-write ourselves into a new history, to be re-born into something greater.

Homo faber need not lay dormant, give up, in the face of collapse.

Conclusion

So we might end by asking the question one more time. Where do *we* begin? Where does *the we* begin?

Notes

Preface

1. Miss Edith Carpenter, "Burning Oil Tank, Titusville, Pa," duplicate postcard, 1915, in author's possession.
2. Williams, *The Long Revolution*, xv.
3. Black, *Crude Oil*, 5–11.
4. Wenzel, introduction, 8.
5. McKibben, *End of Nature*, 40–80.
6. Mouhot, "Past Connections and Present Similarities in Slave Ownership and Fossil Fuel Usage," 329–30.

Introduction. The Mineral Moment

1. For a definition of "fossil capitalism," see chap. 13 in Malm, *Fossil Capital*.
2. Foucault, "Nietzsche, Genealogy, History," 76.
3. For a definition of "slow violence," see chap. 1 in Nixon, *Slow Violence and the Environmentalism of the Poor*.
4. McKibben, *Eaarth*, 2.
5. Michael Watts has documented the type of structural leakage that occurs in sites of oil extraction. See, for example, "A Tale of Two Gulfs," 444.
6. Mark Perry, "New US Homes Today Are 1,000 Square Feet Larger than in 1973 and Living Space per Person Has Nearly Doubled," American Enterprise Institute, June 5, 2016, www.aei.org/publication/new-us-homes-today-are-1000-square-feet-larger-than-in-1973 -and-living-space-per-person-has-nearly-doubled; Christopher Ingraham, "The American Commute Is Worse Today Than It's Ever Been," *Washington Post*, February 22, 2017, www .washingtonpost.com/news/wonk/wp/2017/02/22/the-american-commute-is-worse-today -than-its-ever-been/?utm_term=.5055803d855a; Office of Disease Prevention and Health Promotion, "Estimated Calorie Needs Per Day," *Dietary Guidelines, 2015–2020*, https://health .gov/dietaryguidelines/2015/guidelines/appendix-2.
7. Malm defines the "fossil economy" as "an economy of self-sustaining growth predicated on the growing consumption of fossil fuels, and therefore generating a sustained growth in emissions of carbon dioxide" (11–13, 319–20).
8. White, "Energy and the Evolution of Culture," 335–56; Sieferle, *The Subterranean Forest*; Wrigley, *Continuity, Chance, Change*; Pomeranz, *The Great Divergence*; Catton, *Overshoot*; McNeill, *Something New under the Sun*.
9. See Malm, chaps. 1 and 9.
10. Ibid., 1–19, 194–222.
11. A number of scholars have sketched out a detailed, and eye-opening, portrait of some of the major players and developments in the history of fossil capitalism in the United

States, including Andrews, *Killing for Coal*, Jones, *Routes of Power*, LeCain, *Mass Destruction*, and Needham, *Power Lines*.

12. Heede, "Tracing Anthropogenic Carbon Dioxide and Methane Emissions to Fossil Fuel and Cement Producers." See also the Carbon Majors website (http://carbonmajors.org) for recent information on the major contributors to climate-changing emissions.

13. Adas, *Machines as the Measure of Man*.

14. Huber, *Lifeblood*, 155–70.

Chapter 1. Mineral Rites

1. Foucault, *Discipline and Punish*, 25–28.

2. The US Energy Information Administration (EIA) estimates the 2010 global energy consumption at 524 quadrillion BTU, of which approximately 84% was made up of fossil fuels, or the equivalent of approximately 16 billion tons of coal. In the preindustrial world, an acre of forest could sustainably produce about .73 tons of coal equivalent, and thus we might grossly think of our fossil fuel consumption as something like the equivalent of 21.7 billion acres of forest lands ("International Energy Outlook 2013," 1, www.eia.gov/outlooks /ieo/pdf/0484(2013).pdf). For an estimate of the energy yield of a premodern forest, see Vaclav Smil, quoted in Pomeranz, *The Great Divergence*, 308–9.

3. Skocz, "Husserl's Coal-Fired Phenomenology," 18.

4. Ibid., 20.

5. Quoted in Bachelard, *The Poetics of Space*, 91.

6. Ibid., 3–37.

7. Johnson, *Carbon Nation*, 21–26.

8. Matthew Huber offers a portrait of how the work of cracking hydrocarbons into various fuels and products produced what he calls the "fractionated lives" of the post–WWII period in which the American standard of living became deeply rooted in oil and its products, which, he suggests, played a material role in propping up an entrepreneurial ethos of freedom and mobility that aligns symbolically with the flexibility of a plasticized world (*Lifeblood*, 302–7).

9. Bachelard, 57.

10. Nikiforuk, *The Energy of Slaves*, 71.

11. Johnson, 14–21.

12. EIA, "How Much Energy Does a Person Use in a Year?", June 12, 2018, www.eia.gov /tools/faqs/faq.cfm?id=85&t=1. The figure three hundred million BTUs per capita computes to approximately 10.8 tons of coal equivalent.

13. Needham's *Power Lines* develops in great detail this relationship between the Navajo and the making of the American Southwest. It exemplifies neatly the dialectic between class privilege and subaltern injury, economic growth and its externalities. See chapters 4 and 5, 123–84.

14. EPA, "EPA Cuts Emissions at Navajo Generating Station," July 28, 2014, http://yose mite.epa.gov/opa/admpress.nsf/docf6618525a9efb85257359003fb69d/e26806c557e820e48 5257d2300664dfa!OpenDocument.

15. See Jordan Schneider, Travis Madsen, and Julian Boggs, *America's Dirtiest Power Plants: Their Oversized Contribution to Global Warming and What We Can Do about It*, Envi-

ronment America Research and Policy Center, September 2013, table A2, 28, https://environ
mentamericacenter.org/sites/environment/files/reports/Dirty%20Power%20Plants.pdf,
and EIA, "Energy Mapping System," https://www.eia.gov/state/maps.cfm.

16. See, for instance, EPA, "Addressing Uranium Contamination on the Navajo Reservation," www.epa.gov/Navajo-nation-uranium-cleanup, and Voyles, *Wastelanding*.

17. Navajo National Department of Justice, "Statement of the Navajo Nation Department of Justice to the Office of High Commissioner for Human Rights Universal Periodic
Review of United States of America," 2010, http://lib.ohchr.org/HRBodies/UPR/Documents
/session9/US/NAVAJO_NavajoNationalDepartmentofJustice.pdf.

18. City News Service, "LADWP Owns Big Hunk of Arizona Coal-fired Power Plant and
Is Ready to Sell," *Los Angeles Daily News*, June 22, 2015, www.dailynews.com/business
/20150622/ladwp-owns-big-hunk-of-arizona-coal-fired-power-plant-and-is-ready-to-sell;
Molly Peterson, "Los Angeles DWP Commissioners vote to sell holdings in coal-fired Navajo Generating Plant," 89.3KPCC, May 19, 2015, www.scpr.org/news/2015/05/19/51809/los
-angeles-dwp-commissioners-vote-to-sell-holding.

19. Navajo Nation Economic Development, "An Overview of the Navajo Nation—
Demographics," http://navajobusiness.com/fastFacts/demographics.htm.

20. "Working with Navajo Nation for Cleaner, Healthier Heat," *EPA Science Matters
Newsletter*, October 2014, www.epa.gov/sciencematters/epa-science-matters-newsletter
-working-navajo-nation-cleaner-healthier-heat.

21. Needham, 182.

22. On "energy deepening," see Diamanti, "Aesthetic Economies of Growth."

23. Katie Hunt and Shen Lu, "Smog in China Closes Schools and Construction Sites, Cuts
Traffic in Beijing," CNN, December 8, 2015, www.cnn.com/2015/12/07/asia/china-beijing
-pollution-red-alert.

24. Agence France-Presse, "Beijing's Smog 'Red Alert' Enters Third Day as Toxic Haze
Shrouds City," *Guardian*, December 21, 2015, www.theguardian.com/world/2015/dec/21/
beijings-smog-red-alert-enters-third-day-as-toxic-haze-shrouds-city.

25. For steel, see World Steel Association, "Total Production of Crude Steel Production,
2006–2015," table 1, www.worldsteel.org/en/dam/jcr:37ad1117-fefc-4df3-b84f-6295478ae460
/Steel+Statistical+Yearbook+2016.pdf. For estimates of cement, see United States Geological Survey, "Cement Statistics and Information: Mineral Commodities Summary," 39,
http://minerals.usgs.gov/minerals/pubs/commodity/cement/mcs-2015-cemen.pdf. For estimates of glass and glassware exports, see United Nations, Comtrade Databank, http://com
trade.un.org/db/ce/ceSnapshot.aspx?px=H1&cc=70, which shows that in 2015, China was
the world's largest glass exporter by value and the United States was the world's largest
importer.

26. World Health Organization, "Ambient Outdoor Air Quality and Health," May 2,
2018, www.who.int/mediacentre/factsheets/fs313/en.

27. Edward Wong, "In China, Breathing Becomes Childhood Risk," April 23, 2013, www
.nytimes.com/2013/04/23/world/asia/pollution-is-radically-changing-childhood-in
-chinas-cities.html.

28. Sona Patel, "In the Dirtiest Cities, Air Pollution Forces Life Changes," *New York
Times*, December 22, 2015, www.nytimes.com/2015/12/23/world/asia/air-pollution-china

-india.html?mtrref=undefined&login=email. For an example of recent environmental health risks related to glass manufacturing, including high levels of cadmium and arsenic, in the green city of Portland, Oregon, see Kirk Johnson, "Toxic Moss in Portland, Ore., Shakes City's Green Ideals," *New York Times*, March 2, 2016, www.nytimes.com/2016/03/03/us/toxic -moss-in-oregon-upsets-city-known-for-environmental-ideals.html.

29. Ryan Peterson, "America's Yoga Imports Are Soaring," *Business Insider*, January 18, 2013, www.businessinsider.com/americas-yoga-imports-are-soaring-2013–1.

30. Adela Lin, "Taiwan Blames Chemicals Company for Deadly Kaohsiung Explosions," *Bloomberg Business*, August 2, 2014, www.bloomberg.com/news/articles/2014–08–02/res cuers-seek-survivors-of-taiwan-blasts-that-left-27-dead; Ralph Jennings, "Deadly Taiwan Explosions Expose 'Relaxed' Land Use Rules," *Forbes*, August 7, 2014, www.forbes.com/sites /ralphjennings/2014/08/07/taiwan-confronts-relaxed-land-use-rules-after-deadly-gas -explosions/#205ccda733e0.

31. John Buccini, *The Global Pursuit of the Sound Management of Chemicals*, World Bank, February 24, 2004, 9, http://documents.worldbank.org/curated/en/616191468137102069 /pdf/451290WP0Box331emicals200401PUBLIC1.pdf.

32. Daniel Workman, "Taiwan's Top Ten Exports," World's Top Exports (WTEx), February 15, 2016, www.worldstopexports.com/taiwans-top-exports.

33. Huang et al., "Bisphenol A (BPA) in China," 7.

34. Taiwan Environmental Protection Agency, "Formosa Plastics Renwu Plant to Be Announced Soil Pollution Remediation Site." For a more recent study on water contamination, linked to Formosa Plastics and other producers in this industrial corridor, see Lin, Mao, and Nadim, "Forensic Investigation of BTEX Contamination in Houjing River in Southern Taiwan," 395–402.

35. Huang et al., 2–7.

36. The global scope of environmental health costs of Formosa Plastics include the ironically named Point Comfort, Texas, where one of the company's other subsidiaries released a host of carcinogenic VOCs into the groundwater and soil before reaching a settlement in 2012, and the rural community of Illiopolis, Illinois, where a 2004 vinyl chloride explosion at one of its subsidiaries killed 5 workers and forced 150 people to evacuate to escape toxic fumes. See, for example, EPA, "Case Summary: Settlement with Formosa Plastics Corporation for Site-wide Corrective Actions at Point Comfort, Texas Facility," June 14, 2012, www .epa.gov/enforcement/case-summary-settlement-formosa-plastics-corporation-site-wide -corrective-actions-point.

37. Mark Kinver, "Accumulating Microplastic Threat to Shores," BBC News, January 27, 2012, www.bbc.com/news/science-environment-16709045.

38. See EPA, "Summary of Criminal Prosecutions," 2012, and Department of Justice press release, "Louisiana Oil Refinery Vice-President Pleads Guilty to Air Pollution Causing Negligent Endangerment," July 6, 2011, both available at https://cfpub.epa.gov/compliance /criminal_prosecution/index.cfm?action=3&prosecution_summary_id=2239.

39. Patti Domm, "US Becoming 'Refiner to the World' as Diesel Demand Grows," NBC News, August 7, 2013, www.nbcnews.com/business/us-becoming-refiner-world-diesel-de mand-grows-6C10867716.

40. EIA, "Number and Capacity of Operable Petroleum Refineries by PAD District and State as of January 1, 2015," www.eia.gov/petroleum/refinerycapacity/table1.pdf.

41. Watts, "Oil Frontiers," 190.

42. EPA, "Summary of Criminal Prosecutions."

43. Allen, *Uneasy Alchemy*, 1; Misrach and Orff, *Petrochemical America*.

Chapter 2. Carbon's Social History

1. See Arjun Appadurai's description of the social life of commodities in *The Social Life of Things*, 3–63.

2. Szeman, "How to Know about Oil," 146.

3. This argument is developed in Burke, "The Big Story," and Johnson, *Carbon Nation*.

4. De Kerbrech, *Down amongst the Black Gang*, 32–33, 67

5. Hutchings and De Kerbrech, *RMS* Titanic *Manual*.

6. Ibid., 73–74.

7. Morgan, *Rebirth of a Nation*, 125; Smith, 160–61.

8. Quoted in LeMenager, *Living Oil*, 3.

9. This is taken from the broader portrait of the Welsh coal industry given by Chris William in *Capitalism, Community, and Conflict*. Conditions, of course, varied to an extent from place to place.

10. Collier Killed at Six Bells," *South Wales Echo*, September 28, 1899, 3; "Man Killed at Abertillery," *Evening Express* (Cardiff, Wales), January 11, 1902, 3; "Inquest at Abertillery," *Evening Express* (Cardiff, Wales), September 8, 1903, 3; "Fatality at Trebarris," *Evening Express* (Cardiff, Wales), May 24, 1904, 2; "Fall of Roof," *Evening Express* (Cardiff, Wales), November 24, 1904, 3; "Six Bells Colliery Fatality," *Evening Express* (Cardiff, Wales), February 10, 1906, 2.

11. "Perils of the Pit," *Weekly Mail* (Cardiff, Wales), June 27, 1908, 8; "At Lancaster's Six Bells," *Cardiff Times*, July 27, 1907, 7; "Fatal Fall of Roof," *Evening Express* (Cardiff, Wales), September 29, 1909, 2.

12. "Miners and the L.R.C.," *Evening Express* (Cardiff, Wales), May 15, 1908, 3.

13. "Masters and Workmen," *Evening Press* (Cardiff, Wales), December 16, 1902, 3.

14. For accounts of the riots, see Smith, 162–63, and "Coal Strike Riots," *Cardiff Times and South Wales Weekly News*, November 12, 1910, 8.

15. Smith, 169–70.

16. "Statue Commemorates Six Bells Colliery Disaster," BBC, June 28, 1910, http://news.bbc.co.uk/local/southeastwales/hi/people_and_places/history/newsid_8754000/8754198.stm.

17. De Kerbrech, 41.

18. Clark and Clark, "The International Mercantile Marine Company," 139–142.

19. Barczewski, *Titanic*, 257.

20. De Kerbrech, 41–42. The term *cannibalize* comes from Barczewski, 257. Beyond these sources, there is at least one apocryphal acquisition that makes claims on the voyage's thrust. Not long ago a revolver from one of the Titanic's bursars, a man named George Bull, was put up at public auction for £200,000; the story is that Bull and another agent had

traveled to Wallasey in Merseyside in advance of the Titanic's voyage to purchase coal at gunpoint from local miners. We have only oral history to confirm that story, but the class hostility that it invokes—with middle-class guns pulled on striking miners—speaks to the implied structural violence that gave to the *Titanic*—and that still gives to the modern world—its dynamism.

21. De Kerbrech, 41.

22. Ibid., 22–30.

23. Ibid. Also, the social architecture of the *Titanic*, including the location of the black gang's quarters, the firemen's passage, and so forth are laid out in Hutchings and De Kerbrech's *RMS* Titanic *Manual*.

24. Hutchings and De Kerbrech, 86.

25. De Kerbrech, 65–71.

26. Beesley, *The Loss of the SS* Titanic, 17. This book was originally published in 1912.

27. Brewster, *Gilded Lives, Fatal Voyage*, 77.

28. Ibid., 5–121.

29. Scranton, *Proprietary Capitalism*, 198–99; Hall, *America's Successful Men of Affairs*, 256–57.

30. Brewster, 32–33.

31. Ibid., 34, 58.

32. For a detailed description of the ship's facilities and its material culture, see Hutchings and De Kerbrech, and Gill, *Titanic*.

33. Barczewski, 258.

34. Brewster, 107–8. The *Titanica* exhibit at the Ulster Folk and Transport Museum refers to "electric baths" and "electric beds," for example; see https://nmni.com/titanic/On -Board/Activities-on-board/1st-Class-Turkish-Baths.aspx.

35. Barczewski, 21.

36. A contemporaneous description of using this electrical equipment can be found in Onken and Baker, *Harper's How to Understand Electrical Work*, 232. Carolyn de la Pena has written the most comprehensive recent account of the discourses of health relating to electricity, mechanical devices, and the body in *The Body Electric*.

37. Gill, 140. For a sampling of its provisions, see Barczewski, 257–58.

38. Huber argues in *Lifeblood* that oil provided the material foundation for the expansion in the post–WWII period of American beliefs in the entrepreneurial life and bourgeois individualism. To an extent his argument is applicable to the period before the war; an embrace of entrepreneurship and individualism was evident in more restricted ways in the life patterns derived from coal veins.

39. Descriptions of the first-, second-, and third-class cabins can be found in the *Titanica* exhibit at the Ulster Transport and Folk Museum website, https://nmni.com/titanic/On -Board/Sleeping/3rd-Class-Two-Berth-Stateroom.aspx.

40. Ibid.

41. Brewster, 12.

42. Behe, *On Board the RMS* Titanic, 111; Gleicher, *The Rescue of the Third Class on the* Titanic, 41–44.

43. Gleicher, xii–xv.

44. Nick Barratt details the sequence of events in *Lost Voices from the* Titanic, 110–14.

45. Ibid., 145–46.

46. De Kerbrech, 132–41; Barratt, 126–30.

47. De Kerbrech, 132–41.

48. Barratt, 142, 152.

49. Gleicher details the survival rates by class and ethnicity and argues that a number of decisions were made to maintain order based on ethnic and class assumptions of steerage men being "dangerous individuals" (41–44, 277–78). This assessment also fits with Steven Biel's contention in *Down with the Old Canoe* that racial and class prejudices fundamentally structured how the disaster was represented in public memory (42).

50. Gleicher, 41–44, 277–78.

51. Helen Churchill Candee, "Sealed Orders," *Collier's Weekly*, May 4, 1912, 10–13.

52. Behe, 134, 391. See also Biel, who argues that gender and class myths of civility and danger have framed how this disaster was remembered after the fact as survivors struggled to make sense of acute trauma (16, 27, 29).

53. Barratt, 164.

Chapter 3. Energy Slaves

1. For a discussion of technofundamentalism, see Dinerstein, "Technology and Its Discontents," 569. Andrew Hoffman offers a lucid interpretation of what he calls the "optimistic" path that we are hardwired to endorse in thinking through technology and climate change (*How Culture Shapes the Climate Change Debate*, 30).

2. Petrocultures Research Group, *After Oil*, 13.

3. The term is from the *San Antonio Light*, September 6, 1931. For additional accounts, see E. C. Taylor, "Machines That Are Almost Human: Mechanical Men," *Buffalo (IA) Tribune*, April 19, 1931, 6, "Wonders of Electricity," *Baldur (Manitoba) Gazette*, July 2, 1931, 8, "Armory Prepared for Progress Exposition Which Opens Monday," *Syracuse (NY) Herald*, May 3, 1935, 3; "An Electric Flea Circus," *Mail* (Adelaide, South Australia), January 3, 1931, 19, Schaut, *Robots of Westinghouse*, Cybernetic Zoo, www.cyberneticzoo.com, and Paleo Future, www.paleofuture.com.

4. On the anthropomorphizing of Westinghouse robots, see Toton, "From Mechanical Men to Cybernetic Skin Jobs," 30–31, 54.

5. "An Electric Flea Circus," 19.

6. "Higbie, "Why Do Robots Rebel?"

7. Taylor, 6; Toton, 69–74. See also Gregory Hampton's *Imagining Slaves and Robots in Literature, Film, and Popular Culture* for a discussion of discursive parallels between antebellum representations of slavery and robotic narratives of servitude.

8. Pomeranz, *The Great Divergence*.

9. Kakoudaki, *Anatomy of a Robot*, 161.

10. For background, see Rabinbach, *The Human Motor*, 49–52.

11. Johnson, *Carbon Nation*, 42–46.

12. House of Commons, "Slaves," 25.

13. Ibid.

14. Ibid., 28.

15. de Vogüé, "Electricity at the Paris Exposition," 193.

16. Ibid.

17. Ibid., 195–97. E. A. Wrigley notes that economist Pierre Emile Levasseur also developed this analogy in the 1880s (*Continuity, Chance, Change*, 76).

18. Quoted in Dinerstein, 569.

19. Pennsylvania Joint Committee on Electrification, *Rural Electrification in Pennsylvania*, 4; *Power and the Land*, directed by Joris Ivens (Rural Electrification Administration, 1940), VHS.

20. Gainaday Company advertisement, *Good House Keeping*, February 1919, 152.

21. "About Reddy Kilowatt," www.reddykilowatt.org/about.

22. US Census Bureau, "Lighting Equipment: Housing—General Characteristics," 1940, table 8a, 24.

23. US Census Bureau, "Median Value of Assets for Households," 2011, table 1, https://www.census.gov/data/tables/2011/demo/wealth/wealth-asset-ownership.html.

24. International Energy Agency, "World Energy Outlook 2017," executive summary, 6, https://iea.org/textbase/npsum/weo2017sum.pdf.

25. Jackson, "Idle Slaves of the South," 613–14, 655–57.

26. Ibid, 613–14.

27. Ibid, 655–57.

28. "Thirteen Slaves for a Nickel," 552–53.

29. Ibid.

30. Ibid.

31. According to Toton, the concept of the robot as servant was tied to a gendered logic of marketing domestic help to middle-class housewives (39, 56–57).

32. Rabinbach, 64–96.

33. "Proceedings of the Committee on Dynamics Held at the Franklin Institute," 6.

34. Smith, "Manpower Plus Horsepower," 29–30.

35. Gilbert and Pogue, *Power*, 7.

36. Fuller, "U.S. Industrialization," 57.

37. White, "Energy and the Evolution of Culture," 335–40.

38. Ibid.

39. McNeill, *Something New under the Sun*, 11–12, 15–16.

40. See Pritchard, "Situating Routes of Power within the History of Technology," 20, and Jones, "Response," 28–30.

41. Nikiforuk, *The Energy of Slaves*, 65.

42. LeMenager, "Comments," 12.

43. Rönnbäck, "Slave Ownership and Fossil Fuel Usage," 6.

44. Degani, Hornborg, Strauss, and Love, "Theorizing Energy and Culture," 74.

45. Nikiforuk, 65.

46. Mouhot, "Past Connections and Present Similarities in Slave Ownership and Fossil Fuel Usage," 329–30.

47. Debeir, Deléage, and Hémery, *In the Servitude of Power*, 36, 60; Nikiforuk, 3. The emphasis is mine.

48. Nikiforuk, 4–6, 13–14.

49. Jones, 28–30.

50. Hornborg, "The Fossil Interlude," 50.

51. Johnson, 3–40; Huber, *Lifeblood*.

52. de Jouvenel, "Utopia for Practical Purposes," 442–43.

53. Pritchard, 5.

Chapter 4. Fossilized Mobility

1. Solnit, *Wanderlust*, 9; Kern, *The Culture of Time and Space*.

2. *Automobility* is a term developed in the writings of a number of automobile critics. Typically, it is used to define the culture of the automobile itself, but I have broadened it, with, I think, good reason, to mean the reliance on the various technologies of automobility from the railway and trolley to the automobile and airplane, each of which permits the body to move without labor. The risk here is that the unique cultures that attend each of these transportation technologies over time and space gets diminished some, but the advantage is that such a broadening highlights the immense cultural and material divide between this modern culture of movement and the one that preceded it. See Seiler, *Republic of Drivers*.

3. Associated Press, "Americans Drive 3.1 Trillion Miles in 2015, a New Record," *Los Angeles Times*, February 22, 2016, www.latimes.com/business/autos/la-fi-hy-driving-record -miles-20160223-story.html; EIA, "Oil: Crude and Petroleum Products Explained," November 28, 2016, www.eia.gov/energyexplained/index.cfm?page=oil_use. EIA, "Passenger Travel Accounts for Most of the World Transportation Energy Use," November 19, 2015, www.eia .gov/todayinenergy/detail.php?id=23832. The figure of 60% derives from the addition of aviation, road travel, sea navigation, and railway (International Energy Agency, "Total Final Consumption by Sector," *Key World Energy Statistics*, 39, https://webstore.iea.org/key-world -energy-statistics-2017).

4. More remarkable is the fact that these figures *underrepresent* the degree to which our expenditure of fossil fuels is tasked with the circulation of bodies, resources, and goods. Not included in that calculation is the energy needed to produce the infrastructure itself that makes all of this movement possible, including such things as the amount of energy spent on molting iron into steel for chassis and engines, on crushing, heating, and pulverizing rock to make concrete roads, and on providing the resource inputs to make asphalt pavement and plastic automotive interiors. Three tons of concrete alone are produced each year per person globally for things like road and bridge building, and the process of its manufacturing generates as much as 5% of the world's emissions, making it arguably the dirtiest and most energy intensive of modernity's industries (World Business Council for Sustainable Development, *The Cement Sustainability Initiative*, July 2002, 13, https://web .archive.org/web/20070714085318/http://www.wbcsd.org/DocRoot/1IBetslPgkEie83rTaoJ /cement-action-plan.pdf).

5. Dwight D. Eisenhower to Chief Motor Transport Corps., November 3, 1919; William C. Greany, "Principal [sic] Facts Concerning the First Transcontinental Army Motor Transport Expedition, Washington to San Francisco, July 7 to September 6, 1919," undated, www .eisenhower.archives.gov/research/online_documents/1919_convoy/principal_facts.pdf.

6. Greany.

7. Journals of Meriwether Lewis and William Clark, Clark, undated, ca. January 21, 1804. All subsequent citations to entries in the journals of Lewis and Clark refer to the University of Nebraska-Lincoln's digital archives of the Corps of Discovery Expedition, https://lewisandclarkjournals.unl.edu/journals.

8. Thomas Jefferson laid out the imperial rationale behind the Corps of Discovery in his confidential letter to Congress on January 18, 1803 (US House of Representatives, "President Thomas Jefferson's Confidential Message to Congress Concerning Relations with the Indians and Proposing an Expedition to Explore across the Continent to the Western Ocean," US National Archives and Records Administration, www.archives.gov/education/lessons/lewis-clark/images/jefferson-letter-04.gif).

9. Cronon, Nature's Metropolis, 63.

10. The term derealization comes from Margaret Morse and is quoted in Seiler, 139.

11. As examples, see the following entries: Lewis, September 17, 1804; Clark, April 7, 1806; Lewis, July 10, 1806; John Ordway, May 5, 1805.

12. Lewis, June 1, 1806; Lewis, June 1, 1805; Patrick Gass, October 17, 1805; Joseph Whitehouse, July 23, 1805.

13. Clark, June 23, 1805.

14. Lewis, April 24, 1805.

15. Lewis, undated, winter 1804-5.

16. Whitehouse, May 29, 1805; Whitehouse and Clark, October 18, 1805; Lewis, April 14, 1805; Clark, January 22, 1806; Clark, January 22, 1806.

17. Clark, October 30, 1805.

18. Clark, November 12, 1805.

19. See, for instance, Ordway, August 17, 1805; Clark, "Weather," September 1806, and Lewis, July 11, 1806.

20. Clark, December 29, 1805.

21. Kern, 123-24, 145.

22. See Scott's concept of legibility in Seeing Like a State, 2-3.

23. Lewis, September 5, 1803.

24. Ordway, May 4, 1805.

25. Ordway, July 11, 1805.

26. Clark, March 17, 1806.

27. Clark, July 12, 1804.

28. Whitehouse, February 22, 1805.

29. Clark, December 6, 1805.

30. Clark, July 1, 1804.

31. Clark, June 30, 1804.

32. Whitehouse, June 22, 1804.

33. Johnson, Carbon Nation, 56.

34. Today's biometric apparatuses like the Fitbit are not much better in representing these material facts, even if they inch us toward an understanding of the strangeness of the modern. For instance, I can track that my body moved over fifteen miles today by bike, that along the way I burned about 350 calories of fuel (i.e., energy), and that my work output for

that short and exhausting forty-five-minute period was approximately 145 watts, or one-fifth the capacity of a horse. With a little extrapolation, I can conclude that this middle-aged male body can muster up (for a short period of heavy exertion) about the same amount of work it takes to move a Subaru Outback about a quarter of a mile down a flat road at a good pace. Still, this is a pale dramatization.

35. Lewis, September 1, 1803.

36. Lewis, July 4, 1805; Ordway, August 16, 1804.

37. Clark, August 18, 1804.

38. See Seiler, chaps. 2 and 4.

39. Quoted in Seiler, 146, 140.

40. Baling invoices, codex C, winter 1804–5.

41. Clark, January 31, 1805.

42. Lewis, July 7, 1805.

43. Clark, January 1, 1805.

44. Lewis, June 8, 1806.

45. Clark, April 14, 1804.

46. Clark, October 21, 1805.

47. Clark, September 8, 1806; Clark, August 28, 1805.

48. Whitehouse, October 20, 1805.

49. Lewis, April 27, 1806.

50. Clark, October 21, 1805.

51. Ibid.

52. Clark, September 14, 1805; Whitehouse, September 14, 1805.

53. Clark, July 4, 1804.

54. White, *The Organic Machine*, 3–29.

55. Whitehouse, May 13, 1805.

56. Clark, May 24, 1804.

57. Ordway, September 30, 1804.

58. Lewis, June 2, 1805.

59. Seiler, 69–104; Huber, *Lifeblood*, vii–xxii.

Chapter 5. *Coal TV*

1. Huber, *Lifeblood*, xii.

2. EIA, *EIA Coal Data: A Reference*, February 1995, 17–18, https://books.google.com /books?id=c6fPzHx5UYEC&pg=PP4&source=gbs_selected_pages&sig=ACfU3U1KNzit ICnSApoLMenSFv8TeSCs5g&hl=en#v=onepage&q&f=false; EIA, "How Much of U.S. Carbon Dioxide Emissions Are Associated with Electricity Generation?," www.eia.gov/tools/faqs /faq.cfm?id=77&t=11.

3. Celinda Lake, Daniel Gotoff, Kristel Pondel, Alex Dunn, and Christine Matthews, "Public Attitudes on Mountaintop Removal," Lake Research Partners, Washington, DC, and Bellwether Research and Consulting, Alexandria, VA, 2011, www.lakeresearch.com/news /mtr/MTR%20Slides.pdf.

4. Nathan Joo, Matt Lee-Ashley, and Michael Madowitz, "Fact Sheet: Five Things You Should Know about Powder River Basin Coal Exports," August 19, 2014, Center for Ameri-

can Progress, www.americanprogress.org/issues/green/report/2014/08/19/95820/fact-sheet
-5-things-you-should-know-about-powder-river-basin-coal-exports.

5. Christopher Jones argues that we have a "petromyopia" in the energy humanities; this is a largely true but oversimplified claim. See his "Petromyopia."

6. "Cobalt Coal Reopens Westchester Mine," *Coal Age*, July 20, 2010, accessed February 12, 2017, www.coalage.com/coal-in-the-news/latest/cobalt-coal-re-opens-westchester-mine.

7. Spike TV, "Learn the Truth about Coal," *Spike*, February 25, 2011, www.spike.com /articles/7n8hnx/coal-learn-the-truth-about-coal; Spike TV, "Spike TV Digs 'Coal,'" October 14, 2010, www.spike.com/articles/qi5luw/coal-spike-tv-digs-coal; Mandi Bierly, "'Deadliest Catch' Producer Talks Upcoming Season and New Show 'Coal,'" *Entertainment Weekly*, March 23, 2011, www.ew.com/article/2011/03/23/thom-beers-coal-deadliest-catch.

8. Epstein and Steinberg, "Life in the Bleep Cycle," 91.

9. Ibid.; "'Deadliest Catch' and 'Axe Men' Creator Thom Beers Talks About His New Show, 'COAL'."

10. "'Deadliest Catch' and 'Axe Men' Creator Thom Beers Talks About His New Show, 'COAL'."

11. Dana Jennings, "Grab a Brew while They Face Death," *New York Times*, March 24, 2011, www.nytimes.com/2011/03/27/arts/television/coal-on-spike-aims-to-attract-male -viewers.html?_r=0.

12. "'Deadliest Catch' and 'Axe Men' Creator Thom Beers Talks About His New Show, 'COAL'."

13. Associated Press, "Mine in TV Show 'Coal' Gets Fined for Endangering Miners," April 8, 2011, www.foxnews.com/entertainment/2011/04/08/tv-coal-gets-fined-endanger ing-miners.html.

14. William Egginton proposes that viewers turn to reality television "to watch the human dramas that emerge around those who lose" ("The Best or Worst of Our Nature," 179). While Egginton is speaking about reality contest shows, docudramas like Beers's series likewise, I suggest, trades on a similar pursuit of pain.

15. Spike TV, "Spike TV Digs Down Deep for New Original Series," January 5, 2011, www.prnewswire.com/news-releases/spike-tv-digs-down-deep-for-new-original-series -coal-112943774.html.

16. Bakhtin explains that the novel has a dialogic quality to it in which the reader enters vicariously into different characters' lives and consciousnesses; see *The Dialogic Imagination*, 259–422.

17. Adorno, *The Stars Down to Earth*, 150.

18. The material was perceived to be so good that the scene was apparently restaged for dramatic effect after the initial "real" firing went unrecorded (Zack Harold, "'Coal' Puts on Best Fact," *Charleston (WV) Gazette Mail*, March 30, 2011, www.wvgazettemail.com/News /201103290858).

19. See, for instance, the opening credits, episodes 4, 7, and 9.

20. In episode 4, we hear that "for Mike, keeping the mine afloat is about more than just business. I'm a bigtime family man and, ah, that's why I'm doing this. . . . I'm doing it for the family. We got three kids and a wife that are counting on me. We are not a big corporation. Our necks are on the line." We also learn that Mike relies on G & L Trucking and

that too is a "family business . . . Family man, good father, you see his son hanging with him . . . Those are the kind of people I want to do business with."

21. See the *Coal* blog stream, wvbroadcasting.net, Coolbreeze, April 23, 2011, www.wvbroadcasting.net/viewtopic.php?f=3&t=18765.

22. Trip Gabriel, "50 Years into the War on Poverty, Hardship Hits Back," *New York Times*, April 20, 2014, www.nytimes.com/2014/04/21/us/50-years-into-the-war-on-poverty -hardship-hits-back.html.

23. Lindsey Abrams, "How the Coal Industry Took over West Virginia," *Salon*, March 3, 2014, www.salon.com/2014/03/31/how_the_coal_industry_took_over_west_virginia; Trip Gabriel, "West Virginia Coal Country Sees New Era as Donald Blankenship Is Indicted," November 30, 2014, www.nytimes.com/2014/12/01/us/west-virginia-coal-country-sees-new -era-as-a-mine-boss-is-indicted.html?_r=0.

24. US Census Bureau, "Quick Facts: McDowell County, West Virginia," www.census.gov /quickfacts/table/EDU635214/54047; "County Profile: McDowell County, West Virginia," Institute for Health Metrics and Evaluation, www.healthdata.org/sites/default/files/files /county_profiles/US/2015/County_Report_McDowell_County_West_Virginia.pdf.

25. The Cobalt Mine was a nonunionized mine. Consequently, the difficult subjects of unionization, wages, and benefits were scrupulously left out of the series by the producers. Moreover, not long after the series was canceled in late 2011, the Cobalt miners decided to secure legal counsel and collective bargaining rights under the United Mine Workers Association, suggesting that things were not so, well, let us say, familial down in the pit. The paternal elders in the Cobalt family were running at the time (and on the side) a coal consulting business called New Tech Mining, Inc., that promised they could deliver to would-be investors a "100% union free workforce" as well as much coveted "confidentiality—internally and externally." The Cobalt family might not be so friendly. Crowder's company responded swiftly and brutally by laying off the unionized miners, closing the mine, and then reopening it shortly after with nonunionized labor supplied by a subcontractor. That episode in union busting was so transparent that Cobalt Coal was ordered by the National Labor Relations Board to rehire the workers and to compensate them with back pay. Unfortunately, the story ended when Cobalt did what all good union busters do: it simply closed down the mine; and left the twenty-three miners who were discriminated against with approximately five hundred thousand dollars lost in back pay. See Tom Roberts and Mike Crowder, "New Tech Mining, Inc.: Company Mining Abilities and Startup Support," http://newtechmining .com/NEW%20TECH%20MINING,%20INC%20presentation_files/frame.htm, and Paul Nyden, "Coal Company from Spike TV Series Cited by NLRB," *Charleston (WV) Gazette-Mail*, April 2, 2014, www.wvgazettemail.com/business/coal-company-from-spike-tv-series-cited -by-nlrb/article_159920a4-5b8b-5561-be0e-116829cbc237.html.

26. Ben Jervey, "*Coal*, the Great New Reality TV Show That Totally Misleads You about Modern Mining," *Good*, April 8, 2011, www.good.is/articles/coal-the-great-new-reality-show -that-doesn-t-actually-show-the-reality-of-modern-mining.

27. Halttunen, "Humanitarianism and the Pornography of Pain in Anglo-American Culture."

28. Al Norton, "411mania Interviews: Thom Beers," March 30, 2011, http://411mania .com/movies/411mania-interviews-thom-beers.

29. See Appadurai, introduction to *The Social Life of Things*.

30. Ibid. I have adopted here the language of Appadurai, although I have repurposed it to fit my discussion of the ecology and class politics of this fuel commodity.

31. Bierly.

32. Mandi Bierly, "Spike's Reality Show 'Coal' Earns Violations for Coal Mine," April 8, 2011, http://www.ew.com/article/2011/04/08/spike-coal-mine-violations; Marc Hoffstatter, "Learn 10 Things about Coal in This Episode Three Summary," *Spike*, April 4, 2012, www.spike.com/articles/dxsjtf/coal-learn-10-things-about-coal-in-this-episode-3-summary.

33. Jervey.

34. EIA, "Major US Coal Mines, 2016," www.eia.gov/coal/annual/pdf/table9.pdf.

35. EIA, "Coal Productivity by State and Mine Type, 2016 and 2015," www.eia.gov/coal/annual/pdf/table21.pdf.

36. Katie Valentine, "Scientists Have Now Quantified Mountaintop Removal Mining's Destruction of Appalachia," *Think Progress*, February 11, 2016, http://thinkprogress.org/climate/2016/02/11/3748303/mountaintop-removal-flattening-appalachia.

37. Spike TV, "Learn the Truth about Coal."

38. Hamilton, "Climate Change," 4.

39. Beck, "Living in the World Risk Society."

40. Simon Hefer, "Editorial: If We Take Away Risk, then Capitalism Is Finished," *Telegraph* (London, UK), September 19, 2007, www.telegraph.co.uk/comment/3642776/If-we-take-away-risk-then-capitalism-is-finished.html.

41. Mankiw, *Principles of Economics*, 583.

42. Mark Layton, "Taking Risks to Create Value—It's What Capitalism's All About!," International Risk Management Institute, Inc., September 2007, www.irmi.com/articles/expert-commentary/taking-risks-to-create-value-its-what-capitalisms-all-about.

Chapter 6. Carbon Culture

1. Petrocultures Research Group, *After Oil*, 41.

2. Macdonald, "Oil and World Literature," 7.

3. Ghosh, "Petrofiction," 29–34.

4. Macdonald, 31.

5. Wenzel, introduction, 11.

6. For example, see LeMenager, *Living Oil*, Worden and Barrett, *Oil Culture*, and the experimental essays in Szeman, Wenzel, and Yaeger's *Fueling Culture*.

7. Hawthorne, "The Ambitious Guest," 174–76.

8. Ibid., 175.

9. Sinclair, *The Jungle*, 69.

10. Ibid., 57, 60, 94.

11. EIA, "Estimated Primary Energy Consumption in US, Selected Years, 1635–1945," www.eia.gov/totalenergy/data/annual/showtext.php?t=ptb1601.

12. Ellison, *Invisible Man*, 5–7.

13. Ibid.

14. Wallace, *Infinite Jest*, 3, 15.

15. Melville, "Paradise of Bachelors and Tartarus of Maids," 221, 222, 227.

16. Ibid., 220–21, 226,

17. Lawrence, "Tickets, Please," 290–91.

18. Ibid.

19. Steinbeck, *Grapes of Wrath*, 35.

20. Ibid., 33, 35.

21. Silko, *Ceremony*, 5, 6, 59.

22. Ibid., 25, 236, 239, 240.

23. Ren Xiangkun Clean Coal Experts Convenor et al. "China's Policies for Addressing Climate Change & Efforts to Develop CCUS Technology," www.worldcoal.org/chinas-policies-addressing-climate-change-efforts-develop-ccus-technology; EIA, "US Primary Energy Consumption by Source and Sector, 2016, www.eia.gov/energyexplained.

24. Dickens, *Great Expectations*, 6, 19.

25. Ibid., 10, 13, 33, 97.

26. Dreiser, *Sister Carrie*, 11–12.

27. Ibid., 6, 11, 25, 31.

28. Di Donato, *Christ in Concrete*, 12, 69.

29. Ibid., 4.

30. Ibid., 66–67, 136.

31. McCarthy, *Blood Meridian*, 3, 37, 46.

32. Ibid., 94, 131.

33. Huber, *Lifeblood*, chap. 3.

34. Barthes, *Mythologies*, 97–98.

35. Stowe, *Uncle Tom's Cabin*, 293, 294, 300.

36. Ibid., 94, 121.

37. Hardy, *Far from the Madding Crowd*, 10.

38. Ibid., 115, 262.

39. Pynchon, *The Crying of Lot 49*, 1, 15, 24.

40. Ibid., 24–25.

41. DeLillo, *Underworld*, 110, 388, 515, 625, 825.

42. Ibid., 120, 121, 285, 809, 823.

43. McNeil, *Something New under the Sun*, 212.

44. Smil, "Nitrogen and Human Food Production," 127.

45. Rølvaag, *Giants of the Earth*, 454.

46. Ibid., 58, 126.

47. Norris, *The Pit*, 32, 33, 42, 107, 165.

48. Ibid., 41, 47, 135, 194.

49. Cheever, *Bullet Park*, 21, 92, 101, 118, 233.

50. Ibid., 226.

51. Ibid., 62, 224, 227.

52. Saunders, *The Brief and Frightening Reign of Phil*, 6, 20.

53. Ibid., 20, 21.

54. Davis, *Life in the Iron Mills, and Other Stories*, 11, 12.

55. Ibid., 12, 24, 25, 31.

56. Toomer, *Cane*, 73.

57. Ibid., 144, 151, 160.

58. Mailer, *The Naked and the Dead*, 20, 26, 210.

59. Ibid., 30, 31, 35, 37, 277.

60. Carter, *The Infernal Desire Machines of Doctor Hoffman*, 15, 193, 224.

61. Ibid., 16–17, 107, 206–10, 212.

Epilogue

1. McKibben, *Eaarth*; Kolbert, *The Sixth Extinction*.

2. See the discussion of standard time and private time in Kern, *The Culture of Time and Space*, chap. 1.

3. Ibid.

4. Hallegatte, Bangalore, et al. *Shock Waves*, 2.

5. World Health Organization, "Climate Change and Health: Fact Sheet," June 2016, www.who.int/mediacentre/factsheets/fs266/en.

6. Brian Kahn, "Antarctica CO_2 Hit 400 PPM for First Time in Four Million Years," Climate Central, June 15, 2016, www.climatecentral.org/news/antarctica-co2–400-ppm -million-years-20451.

7. "Meet the Human Ancestor Who Walked Earth 4 Million Years Ago," *Independent* (London, UK), October 1, 2009, www.independent.co.uk/news/science/meet-the-human -ancestor-who-walked-earth-4-million-years-ago-1796386.html.

8. See the Big History Project at www.bighistoryproject.com/about and Christian, *Maps of Time*, 511.

9. Christian, 1.

10. McNeill, foreword, xv.

11. Christian, 11.

12. Christian, 100.

13. Callison, *How Climate Change Comes to Matter*.

14. McNeill and Engelke, *The Great Acceleration*.

15. Ibid., 5–6.

16. Ibid., 207–8.

17. Jason Moore, "Name the System! Anthropocenes and the Capitalocene Alternate," blog, https://jasonwmoore.wordpress.com/2016/10/09/name-the-system-anthropocenes -the-capitalocene-alternative.

18. Moore, *Capitalism in the Web of Life*, 18–22, 77, 119–21, 182–87.

19. Ibid., 92–125.

20. There is an important point to make on this score. It concerns the distinction between early capital and modern capital, a distinction that the Capitalocene argues is not relevant to marking the epoch of climate change. On the one hand, Marx, who is the anchor of Moore's theory, was himself not clear on the meaning of this distinction in his writings. See Johnson, "The Outer Nature of Capital." More important, however, is that Marxist materialism, even this updated version of it, finds itself bundled up in a linguistic confusion over the meaning of energy that at once recognizes and then effaces the thermodynamic equivalencies that made modern capital's project possible. The magic of the fossil system lies in a deep and barely recognized materialism, constituted by an environmental calculus

of functional equivalencies in which (what Moore calls) "the four cheaps" (cheap energy, cheap resources, cheap labor, and cheap food) don't hold up as categories. The big fact is that fossil capital thrives, as did fossil communism and fossil fascism, on a magical alchemy practiced, minute by minute, hour by hour, year by year, in which energy, labor, heat, and matter swap forms without pause or recognition. Today, food mutates in its grip into labor, labor returns to us as food, timber morphs into steel, and coal resurfaces somehow, and inexplicably, in the form of the forests, fiber, and protein needed to provision the world's population. Talking of "the four cheaps" is not wrong: it is just insufficiently materialist on this score, and that sets us up to once again bury the significance of modernity's break with the premodern world. To be clear, fossil capital's work, documented in detail by energy historians and anthropologists, is a quite particular type of ecological praxis, very different from that of premodern capital, and that distinction matters in both our history and prospects for sustainability.

21. Waters, Stafford, Kooyman, and Hills, "Late Pleistocene Horse and Camel Hunting at the Southern Margin of the Ice-Free Corridor," 4263.

22. For an early summary of New World crops and their impact on global development, see Crosby, *The Columbian Exchange*, and for a basic introduction to the role of agriculture in the rise of cities, empires, and power, see Diamond, *Guns, Germs, and Steel*.

23. This account is related in Clendinnen, "Fierce and Unnatural Cruelty," 81–82.

24. Semonin, *American Monster*; Richard Conniff, "Mammoths and Mastodons: All American Monsters," *Smithsonian*, April 2010, www.smithsonianmag.com/science-nature /mammoths-and-mastodons-all-american-monsters-8898672.

25. John Gibbs, "George Washington's Mastodon Tooth," George Washington's Mount Vernon, www.mountvernon.org/digital-encyclopedia/article/george-washingtons-mastodon -tooth.

26. See Worster, *Wealth of Nature*, 14, 214–20.

27. John Ashworth argues that the American Civil War was at root a bourgeois revolution in "Towards a Bourgeois Revolution."

28. Ibid, 204.

29. Worster, *The Dust Bowl*.

30. USDA, "World Crop Production Summary," *World Agricultural Production*, World Agricultural Outlook Board, May 2017, https://apps.fas.usda.gov/psdonline/circulars/pro duction.pdf.

31. The Intergovernmental Panel on Climate Change issued its third, and arguably most catalytic, international summary report in 2001.

32. Christensen, Hewitson, et al., "2007: Regional Climate Projections."

Bibliography

Adas, Michael. *Machines as the Measure of Man: Science, Technology, and Ideologies of Western Dominance*. Ithaca, NY: Cornell University Press, 1989.

Adorno, Theodore. *The Stars Down to Earth*. New York: Routledge, 2002.

Allen, Barbara L. *Uneasy Alchemy: Citizens and Experts in Louisiana's Chemical Corridor Disputes*. Cambridge, MA: MIT Press, 2003.

Andrews, Thomas. *Killing for Coal: America's Deadliest Labor War*. Cambridge, MA: Harvard University Press, 2008.

Appadurai, Arjun. *The Social Life of Things: Commodities in Cultural Perspective*, ed. Arjun Appadurai, 3–63. New York: Cambridge University Press, 1986.

Ashworth, John. "Towards a Bourgeois Revolution? Explaining the American Civil War." *Historical Materialism* 19.4 (2011): 193–205.

Bachelard, Gaston. *The Poetics of Space: The Classic Look at How We Experience Intimate Places*. Trans. John R. Stilgoe. Boston: Beacon Press, 1994.

Bakhtin, Mikhail M. *The Dialogic Imagination: Four Essays*. Ed. Michael Holquist. Trans. Caryl Emerson and Michael Holquist. Austin: University of Texas Press, 1981.

Barczewski, Stephanie. *Titanic: A Night Remembered*. London: Bloomsbury, 2011.

Barnett, Anthony. Foreword to *The Long Revolution*, by Raymond Williams, 7–25. Cardigan, UK: Parthian, 2013.

Barratt, Nick. *Lost Voices from the* Titanic: *The Definitive Oral History*. London: Arrow Books, 2010.

Barthes, Roland. *Mythologies*. Trans. Annette Lavers. New York: Hill and Wang, 2001.

Beck, Ulrich. "Living in the World Risk Society." *Economy and Society* 35.3 (2006): 329–45.

Beesley, Lawrence. *The Loss of the SS* Titanic: *Its Stories and Lessons*. St. Petersburg, FL: Red and Black Publishers, 2008.

Behe, George. *On Board the RMS* Titanic: *Memories of the Maiden Voyage*. Lulu.com, 2011.

Biel, Steven. *Down with the Old Canoe: A Cultural History of the* Titanic *Disaster*. New York: Norton, 1997.

Black, Brian. *Crude Oil: Petroleum in World History*. Lanham, MD: Rowman and Littlefield, 2012.

Brewster, Hugh. *Gilded Lives, Fatal Voyage: The* Titanic's *First-Class Passengers and Their World*. New York: Crown, 2012.

Burke, Edmund, III. "The Big Story: Human History, Energy Regimes, and the Environment." In *The Environment and World History*, ed. Edmund Burke III and Kenneth Pomeranz, 33–53. Berkeley: University of California Press, 2009.

Callison, Candis. *How Climate Change Comes to Matter: The Communal Life of Facts*. Durham, NC: Duke University Press, 2014.

Cane, Jean. *Cane*. New York: Norton, 2011.

Carter, Angela. *The Infernal Desire Machines of Doctor Hoffman*. New York: Penguin, 1994.

Catton, William. *Overshoot: The Ecological Basis of Revolutionary Change*. Urbana: University of Illinois Press, 1980.

Cheever, John. *Bullet Park*. New York: Vintage, 1992.

Christensen, Jens Hesselbjerg, Bruce Hewitson, et al. "2007: Regional Climate Projections." In *Climate Change 2007: The Physical Science Basis. Contribution of Working Group I to the Fourth Assessment Report of the Intergovernmental Panel on Climate Change*, ed. Susan Solomon et al., 889–91. Cambridge: Cambridge University Press, 2007.

Christian, David. *Maps of Time: An Introduction to Big History*. Berkeley: University of California Press, 2005.

Clark, John J. and Margaret G. Clark. "The International Mercantile Marine Company: A Financial Analysis." *American Neptune* 57.2 (1997): 139–42.

Clendinnen, Inga. "Fierce and Unnatural Cruelty: Cortés and the Conquest of Mexico." *Representations* 33 (Winter 1991): 65–100.

Conniff, Richard. "Mammoths and Mastodons: All American Monsters." *Smithsonian*, April 2010, www.smithsonianmag.com/science-nature/mammoths-and-mastodons-all-american-monsters-8898672.

Crosby, Alfred. *The Columbian Exchange: Biological and Cultural Consequences of 1492*. Westport, CT: Greenwood Press, 1972.

Davis, Rebecca Harding. *Life in the Iron Mills, and Other Stories*. New York: Feminist Press, 1993.

Debeir, Jean Claude, Jean-Paul Deléage, and Daniel Hémery. *In the Servitude of Power: Energy and Civilization through the Ages*. Trans. John Barzman. London: Zed Books, 1991.

Degani, Michael, Alf Hornborg, Sarah Strauss, and Thomas Love. "Theorizing Energy and Culture." In *Cultures of Energy: Powers, Practices, Technologies*. Ed. Sarah Strauss, Stephanie Rupp, and Thomas Love, 73–76. Walnut Creek, CA: Left Coast Press, 2013.

de Jouvenel, Bertrand. "Utopia for Practical Purposes." *Daedalus* 94.2 (1965): 437–53.

de Kerbrech, Richard P. *Down amongst the Black Gang: The World and Workplace of RMS Titanic's Stokers*. Briscombe, UK: History Press 2014.

de la Pena, Carolyn. *The Body Electric: How Strange Machines Built the Modern American Body*. New York: New York University Press, 2003.

DeLillo, Don. *Underworld*. New York: Scribner, 1998.

de Vogüé, Eugène-Melchoir. "Electricity at the Paris Exposition." In *Chautauquan: A Monthly Magazine, 10 October 1889 to March 1890*, 193. Meadville, PA: Flood Publishing, 1889–90.

Diamanti, Jeff. "Aesthetic Economies of Growth: Energy, Value, and the Work of Culture after Oil." PhD diss., University of Alberta, 2015.

Diamond, Jared. *Guns, Germs, and Steel: The Fate of Human Societies*. New York: Norton, 1992.

Dickens, Charles. *Great Expectations*. New York: Shine Classics, 2014.

Di Donato, Pietro. *Christ in Concrete*. New York: New American Library, 2004.

Dinerstein, Joel. "Technology and Its Discontents: On the Verge of the Post-Human." *American Quarterly* 58.3 (2006): 569–95.

Dreiser, Theodore. *Sister Carrie*. Mineola, NY: Dover, 2004.

Egginton, William. "The Best or Worst of Our Nature: Reality TV and the Desire for Limitless Change." *Configurations* 15.2 (2007): 171–91.

Ellison, Ralph. *Invisible Man*. New York: Vintage, 1995.

Epstein, Debbie, and Deborah Lynn Steinberg. "Life in the Bleep Cycle: Inventing Id-TV on the Jerry Springer Show." *Discourse* 25.3 (2003): 90–114.

Foucault, Michel. *Discipline and Punish: The Birth of the Prison*, trans. by Alan Sheridan. New York: Vintage, 1995.

———. "Nietzsche, Genealogy, History." In *The Foucault Reader*, ed. Paul Rabinow, 76–100. New York: Pantheon, 1984.

Fuller, Buckminster. "U.S. Industrialization." *Fortune* 21.2 (1941): 50–57.

Ghosh, Amitav. "Petrofiction." *New Republic*, March 1992, 29–34.

Gilbert, Charles C., and Joseph Ezekiel Pogue. *Power: Its Significance and Needs*. Washington DC: Government Printing Office, 1918.

Gill, Anton. Titanic: *Building the World's Most Famous Ship*. Guilford, CT: Lyons Press, 2011.

Gleicher, David. *The Rescue of the Third Class on the* Titanic: *A Revisionist History*. St. John's, Newfoundland: International Maritime Economic History Association, 2006.

Hall, Henry, ed. *America's Successful Men of Affairs: The United States at Large*. Vol. 2. New York: New York Tribune Company, 1896.

Hallegatte, Stéphane, Mook Bangalore, et al. *Shock Waves: Managing the Impacts of Climate Change on Poverty*. Washington, DC: World Bank, 2016.

Halttunen, Karen. "Humanitarianism and the Pornography of Pain in Anglo-American Culture." *American Historical Review* 100.2 (1995): 303–34.

Hamilton, Lawrence. "Climate Change: Partisanship, Understanding, and Public Opinion." *Carsey Institute* 26 (Spring 2011), http://scholars.unh.edu/cgi/viewcontent.cgi?article=1133&context=carsey.

Hampton, Gregory. *Imagining Slaves and Robots in Literature, Film, and Popular Culture*. Lanham, MD: Lexington Books, 2015.

Hardy, Thomas. *Far from the Madding Crowd*. Ware, UK: Wordsworth, 1997.

Hawthorne, Nathaniel. "The Ambitious Guest." In *Hawthorne: Selected Tales and Sketches*, 174–82. New York: Holt, Rinehart, and Winston, 1970.

Heede, Richard. "Tracing Anthropogenic Carbon Dioxide and Methane Emissions to Fossil Fuel and Cement Producers, 1854–2010." *Climatic Change* 122.1–2 (2014): 229–41.

Higbie, Tobias. "Why Do Robots Rebel? The Labor History of a Cultural Icon." *Labor: Studies in Working-Class History of the Americas* 10.1 (2013): 99–121

Hoffman, Andrew. *How Culture Shapes the Climate Change Debate*. Stanford, CA: Stanford University Press, 2015.

Hornborg, Alf. "The Fossil Interlude: Euro-American Power and the Return of the Physiocrats." In *Cultures of Energy: Powers, Practices, Technologies*, ed. Sarah Strauss, Stephanie Rupp, and Thomas Love, 41–59. Walnut Creek, CA: Left Coast Press, 2013.

House of Commons. "Slaves: Berbice and Demerara," In *Slave Trade*, 2 vols. London: HMSO, 1828.

Huang, Y. Q., et al. "Bisphenol A. BPA. in China: A Review of Sources, Environmental Levels, and Potential Human Health Impacts." *Environment International* 42 (July 2012): 91–99.

Huber, Matthew. *Lifeblood: Oil, Freedom, and Forces of Capital*. Minneapolis: University of Minnesota Press, 2013.

Hutchings, David, and Richard P. De Kerbrech. *RMS* Titanic *Manual, 1909–1912, Olympic Class*. Somerset, UK: Zenith Press, 2011.

Jackson, Marion. "Idle Slaves of the South." *Survey Graphic*, March 1, 1924, 613–14, 655–57.

Johnson, Bob. *Carbon Nation: Fossil Fuels in the Making of American Culture*. Lawrence: University Press of Kansas, 2014.

———. "The Outer Nature of Capital: The Thermodynamic and Technological Revolutions in the Writings of Marx." In *The Bloomsbury Companion to Marx*, ed. Andrew Pendakis and Imre Szeman. London: Bloomsbury Press, forthcoming.

Jones, Christopher. "Petromyopia: Oil and the Energy Humanities." *Humanities* 5.2 (2016): 1–10.

———. "Response." *H-Environment Roundtable Reviews* 5.9. (2015): 24–33.

———. *Routes of Power: Energy and Modern America*. Cambridge, MA: Harvard University Press, 2014.

Kakoudaki, Despina. *Anatomy of a Robot: Literature, Cinema, and the Cultural Work of Artificial People*. New Brunswick, NJ: Rutgers University Press, 2014.

Kern, Stephen. *The Culture of Time and Space, 1880–1918*. Cambridge, MA: Harvard University Press, 2003.

Kolbert, Elizabeth. *The Sixth Extinction*. New York: Henry Holt, 2014.

Lawrence, D. H. "Tickets, Please." In *D.H. Lawrence: The Complete Short Stories*, 290–97. Blackthorn Press, 2002.

LeCain, Timothy. *Mass Destruction: The Men and Machines That Wired and Scarred the Planet*. New Brunswick, NJ: Rutgers University Press, 2009.

LeMenager, Stephanie. "Comments." *H-Environment Roundtable Reviews* 5.9. (2015): 11–13.

———. *Living Oil: Petroleum Culture in the American Century*. Oxford: Oxford University Press, 2014.

Lin, Chitsan, Wen-Ming Mao, and Farhad Nadim. "Forensic Investigation of BTEX Contamination in Houjing River in Southern Taiwan." *Journal of Environmental Engineering Management* 17.6 (2007): 395–402.

Macdonald, Graeme. "Oil and World Literature." *American Book Review* 33.3 (2012): 7–31.

Mailer, Norman. *The Naked and the Dead*. New York: Picador, 2000.

Malm, Andreas. *Fossil Capital: The Rise of Steam Power and the Roots of Global Warming*. London: Verso, 2016.

Mankiw, N. Gregory. *Principles of Economics*. 6th ed. Mason, OH: Southwestern, Cengage Learning, 2009.

McCarthy, Cormac. *Blood Meridian, or the Evening Redness in the West*. New York: Vintage, 1985.

McKibben, Bill. *Eaarth: Making Life on a Tough New Planet*. New York: Henry Holt, 2010.

———. *End of Nature*. New York: Random House, 2006.

McNeill, John R. *Something New under the Sun: An Environmental History of the Twentieth-Century World*. New York: Norton, 2000.

McNeill, John R., and Peter Engelke. *The Great Acceleration: An Environmental History of the Anthropocene since 1945*. Cambridge, MA: Harvard University Press, 2016.

McNeill, William. Foreword to *Maps of Time: An Introduction to Big History*, by David Christian, xv–xviii. Berkeley: University of California Press, 2005.

Melville, Herman. "Paradise of Bachelors and Tartarus of Maids." In *Herman Melville: Selected Tales and Poems*, ed. Richard Chase, 206–29. New York: Holt, Rinehart, and Winston, 1950.

Misrach, Richard, and Kate Orff. *Petrochemical America*. New York: Aperture Foundation Books, 2014.

Morgan, Kenneth O. *Rebirth of a Nation: Wales, 1880–1980*. New York: Oxford University Press, 1981.

Moore, Jason. *Capitalism in the Web of Life: Ecology and the Accumulation of Capital*. New York: Verso, 2015.

Mouhot, Jean-François. "Past Connections and Present Similarities in Slave Ownership and Fossil Fuel Usage." *Climatic Change* 105.1–2 (2011): 329–55.

Needham, Andrew. *Power Lines: Energy in the Making of the American Southwest*. Princeton, NJ: Princeton University Press, 2014.

Nikiforuk, Andrew. *The Energy of Slaves: Oil and the New Servitude*. Vancouver, BC: Greystone Books, 2012.

Nixon, Rob. *Slow Violence and the Environmentalism of the Poor*. Cambridge, MA: Harvard University Press, 2011.

Norris, Frank. *The Pit: A Chicago Story*. CreateSpace Independent Publishing Platform, 2014.

Onken, William H., Jr., and Joseph P. Baker. *Harper's How to Understand Electrical Work*. New York: Harper and Brothers, 1907.

Pennsylvania Joint Committee on Electrification. *Rural Electrification in Pennsylvania*. Harrisburg, PA: Pennsylvania Joint Committee on Electrification, 1928.

Petrocultures Research Group. *After Oil*. Edmonton, AB: Petroleum Research Cluster, 2015.

Pomeranz, Kenneth. *The Great Divergence: China, Europe, and the Making of the Modern World Economy*. Princeton, NJ: Princeton University Press, 2001.

Pritchard, Sara B. "Situating *Routes of Power* within the History of Technology." *H-Environment Roundtable Reviews* 5.9 (2015): 14–23.

"Proceedings of the Committee on Dynamics Held at the Franklin Institute." In *Principles of Dynamics*, by John W. Nystrom. Philadelphia, PA: John P. Murphy Printer, 1874.

Pynchon, Thomas. *The Crying of Lot 49*. New York: Harper Collins, 1999.

Rabinbach, Anson. *The Human Motor: Energy, Fatigue, and the Origins of Modernity*. Berkeley: University of California Press, 1992.

Rølvaag, Edvart Ole. *Giants of the Earth: A Saga of the Prairie*. New York: Harper Perennial, 1999.

Rönnbäck, Klas. "Slave Ownership and Fossil Fuel Usage: A Commentary." *Climatic Change* 122.1–2 (2014): 1–9.

Saunders, George. *The Brief and Frightening Reign of Phil*. New York: Berkeley, 2005.

Schaut, Scott. *Robots of Westinghouse, 1924–Today*. Mansfield, OH: Schaut, 2007.

Scott, James C. *Seeing Like a State: How Certain Schemes to Improve the Human Condition Have Failed*. New Haven, CT: Yale University Press, 1998.

Scranton, Philip. *Proprietary Capitalism: The Textile Manufacture at Philadelphia, 1800–1885*. New York: Cambridge University Press, 1983.

Seiler, Cotten. *Republic of Drivers: A Cultural History of Automobility in America*. Chicago: University of Chicago Press, 2008.

Semonin, Paul. *American Monster: How America's First Prehistoric Creature Became a Symbol of National Identity*. New York: New York University Press, 2000.

Sieferle, Rolfe. *The Subterranean Forest: Energy Systems and the Industrial Revolution*. Cambridge, UK: White Horse Press, 2010.

Silko, Leslie Marmon. *Ceremony*. New York: Penguin, 2006.

Sinclair, Upton. *The Jungle*. Mineola, NY: Dover, 2001.

Skocz, Dennis E. "Husserl's Coal-Fired Phenomenology: Energy and Environment in an Age of Whole-House Heat and Air-Conditioning." *Environmental and Architectural Phenomenology Newsletter* 21.2 (2010): 16–21.

Smil, Vaclav. "Nitrogen and Human Food Production: Proteins for Human Diets." *Ambio* 31.2 (2002): 126–31.

Smith, David. "Tonypandy 1910: Definitions of Community." *Past and Present* 87 (May 1980): 158–84.

Smith, George Otis. "Manpower Plus Horsepower: The Multiplication of American Muscle through Machines." *Nation's Business* (July 1921): 29–30.

Solnit, Rebecca. *Wanderlust: A History of Walking*. New York: Penguin, 2001.

Steinbeck, John. *Grapes of Wrath*. New York: Penguin, 2002.

Stowe, Harriet Beecher. *Uncle Tom's Cabin*. Mineola, NY: Dover, 2005.

Szeman, Imre. "How to Know about Oil: Energy Epistemologies and Energy Futures." *Journal of Canadian Studies* 48.3 (2013): 145–68.

Szeman, Imre, Jennifer Wenzel, and Patricia Yaeger, eds. *Fueling Culture: 101 Words for Energy and Environment*. New York: Fordham University Press, 2017.

Taiwan Environmental Protection Agency. "Formosa Plastics Renwu Plant to Be Announced Soil Pollution Remediation Site." *Environmental Policy Monthly* 13.4 (2010): 10–11. www.epa.gov.tw/public/Data/78161194871.pdf.

"Thirteen Slaves for a Nickel." *Popular Mechanics* (April 1939): 552–55; 114A.

Toton, Sarah. "From Mechanical Men to Cybernetic Skin Jobs: A History of Robots in American Popular Culture." PhD diss., Emory University, 2014.

Voyles, Traci Brynne. *Wastelanding: Legacies of Uranium Mining in Navajo Country*. Minneapolis: University of Minnesota Press, 2015.

Wallace, David Foster. *Infinite Jest*. New York: Back Bay Books, 2016.

Waters, Michael R., Thomas W. Stafford Jr., Brian Kooyman, and L. V. Hills. "Late Pleistocene Horse and Camel Hunting at the Southern Margin of the Ice-Free Corridor: Reassessing the Age of Wally's Beach, Canada." *Proceedings of the National Academy of the Sciences of the United States of America* 112.4 (2015): 4263–267.

Watts, Michael. "Oil Frontiers: The Niger Delta and the Gulf of Mexico." In *Oil Culture*, ed. Ross Barrett and Daniel Worden, 189–210. Minneapolis: University of Minnesota Press, 2014.

———. "A Tale of Two Gulfs: Life, Death, and Dispossession along Two Oil Frontiers." *American Quarterly* 64.3 (2012): 437–67.

Wenzel, Jennifer. Introduction to *Fueling Culture: 101 Words for Energy and Environment*, ed. Imre Szeman, Jennifer Wenzel, and Patricia Yaeger, 1–16. New York: Fordham University Press, 2017.

White, Leslie. "Energy and the Evolution of Culture." *American Anthropologist* 45.3 (1943): 335–56.

White, Richard. *The Organic Machine: The Remaking of the Columbia River.* New York: Hill and Wang, 1995.

William, Chris. *Capitalism, Community, and Conflict: The South Wales Coalfield, 1898–1947.* Cardiff, UK: University of Wales Press, 1999.

Williams, Raymond. *The Long Revolution.* Cardigan, Wales: Parthian Books, 2013.

Worden, Daniel, and Ross Barrett, eds. *Oil Culture.* Minneapolis: University of Minnesota Press, 2014.

Worster, Donald. *The Dust Bowl: The Southern Plains in the 1930s.* Oxford: Oxford University Press, 2004.

———. *Wealth of Nature: Environmental History and the Ecological Imagination.* Oxford: Oxford University Press, 1993.

Wrigley, E. A. *Continuity, Chance, Change: The Character of the Industrial Revolution in England.* Cambridge: Cambridge University Press, 1990.

Index

Page numbers in *italics* refer to illustrations.